Meeting SEN

in the Curriculum:

GEOGRAPHY

Other titles in the Meeting Special Needs in the Curriculum Series:

Meeting Special Needs in English
Tim Hurst
1 84312 157 3

Meeting Special Needs in Maths
Brian Sharp
1 84312 158 1

Meeting Special Needs in Science
Carol Holden and Andy Cooke
1 84312 159 X

Meeting Special Needs in ICT
Michael North and Sally McKeown
1 84312 160 3

Meeting Special Needs in Art
Kim Earle and Gill Curry
1 84312 161 1

Meeting Special Needs in History
Richard Harris and Ian Luff
1 84312 163 8

Meeting Special Needs in PE and Sport
Crispin Andrews
1 84312 164 6

Meeting Special Needs in Modern Foreign Languages
Sally McKeown
1 84312 165 4

Meeting Special Needs in Design and Technology
Louise T. Davies
1 84312 166 2

Meeting Special Needs in Religious Education
Dilwyn Hunt
1 84312 167 0

Meeting Special Needs in Music
Victoria Jaquiss and Diane Paterson
1 84312 168 9

Meeting Special Needs in Citizenship
Alan Combes
1 84312 169 7

Meeting SEN in the Curriculum: GEOGRAPHY

Diane Swift

David Fulton Publishers

David Fulton Publishers Ltd
The Chiswick Centre, 414 Chiswick High Road, London W4 5TF

www.fultonpublishers.co.uk

David Fulton Publishers is a division of Granada Learning Limited, part of ITV plc.

British Library Cataloguing in Publication Data
A catalogue record for this book is available from the British Library.

ISBN: 1 84312 162 X

10 9 8 7 6 5 4 3 2 1

Typeset by Servis Filmsetting Ltd, Manchester
Printed and bound in Great Britain by Ashford Colour Press

Contents

Foreword

Reading this amazing book reminded me of several encounters in my school teaching career. When I started teaching, in the mid 1970s, comprehensive schools were new and the school leaving age had only just been increased to 16 years. There was no National Curriculum. There was very little Continuting Professional Development. One's classroom was 'a black box': what went on in there, both good and bad, was concealed from view.

We now work in classrooms with 'open doors'. Teachers are much more accountable, but they also have far more support and research evidence from which to draw. We have come a long way; and this book shows that very well.

So to those encounters. What did I do with that boy who could not see that his map of the British Isles was back to front? I was sympathetic but I don't think I was much help to him. And did I consider carefully enough the needs of the little girl with brittle bones when I was planning the fieldwork? What about the boy who found numbers so confusing that he couldn't tell the time? I allowed him to avoid those aspects of geography he found difficult, but could I have done something to help him overcome his difficulties? (He went on to win a national essay competition for sixth formers at the age of 14 years.) I was so ill-prepared, and understood little of some of the conditions that differentiate people. Without understanding, you do feel a bit helpless.

I think geography teachers have been coping with these issues but, like me, only just! We have been crying out for a book like this. *Meeting SEN in the Curriculum: Geography* puts accessible information between two covers, enabling us to understand various human conditions better and providing a wealth of practical suggestions about how to respond positively and productively to students' different learning needs.

I am proud of the fact that the Geographical Association, through one of its projects and the tireless leadership of Di Swift, is associated with this work.

Dr David Lambert
Chief Executive, Geographical Association

Acknowledgements

Meeting SEN in the Curriculum: Geography is one outcome of the Geographical Association's Valuing Places project, funded by DfID (Department for International Development) and the Tubney Foundation. This project has enabled groups of teachers to meet and to construct curriculum thinking informed by their own circumstances. The Valuing SEN in Geography team have all contributed to this publication, and many teachers involved in the project have been able to comment on earlier drafts of the materials. We are grateful for the support of this project and for the work of the colleagues listed below:

Dr David Lambert, Chief Executive of the Geographical Association, for his comments and advice;

Staff and students at our schools: Moorside High School, Werrington; St. John Fisher Catholic High School, Newcastle under Lyme; Stretton Community Special Schools Federation, Burton-upon-Trent; and Springfield Community School, Leek;

The Staffordshire Best Practice Research Scholarship teachers led by Peter Davies, Staffordshire University, for the writing scaffolds on pages 84–87 in Chapter 4, www.sln.org.uk/geography.BPRS.htm;

Roger Carter, Pat Crook, Chris Durbin, Kate Russell, Janine Murphy and Julie Stevenson, all of whom have supported Geography and Special Needs Education through their advisory and inspection work with Staffordshire Local Education Authority;

The Geographical Association, and Peter Wright at Birmingham City Council's Department of Planning and Architecture for supplying some of the images used in this book.

Contributors

This text has been written by the Valuing SEN in Geography Team. It has formed part of their work for the Geographical Association's Valuing Places Project (www.geography.org.uk/vp). Their work has been funded by the Tubney Foundation and supported by the other project groups that are funded by the Department for International Development. Diane Swift is the project co-ordinator and leads the Valuing SEN team.

Irene Corden
Irene is currently headteacher at Springfield Community School, Leek. Irene's experience is predominantly in schools catering for students with severe and profound and multiple learning difficulties. Along with Diane Swift she developed the Staffordshire Expanding Geography Scheme.

Jan Bird
Jan is deputy headteacher at Moorside High School, Werrington, 11–18 comprehensive. One of her responsibilities is as SENCO, and she has recently completed an MA at the University of Leicester to help her manage this area.

Julie Dale
Julie is a Geography teacher and SENCO at St. John Fisher Catholic High School in Newcastle Under Lyme, Staffordshire. She is a member of the Best Practice Research Team that developed the writing scaffolds featured in Chapter 4.

Jan Fielding
Jan is a Senior Teaching Assistant at Moorside High School, Werrington. She has particular responsibility for students with Dyslexia (SpLD) and is a Learning Mentor with the Aim Higher Project.

Bev Rowley
Bev is Geography Co-ordinator at Stretton Community Special Schools Federation, Burton-upon-Trent. She is based at Stretton Brook, the secondary phase school for students with moderate and severe learning difficulties. Bev started her career as an NNEB. After completing a degree with the Open University she obtained qualified teacher status through the graduate teacher training programme.

Diane Swift
Diane started her career teaching geography in Birmingham and Walsall comprehensive schools. After running Darlaston Community School's large geography department she was appointed adviser for geography in Staffordshire. It was here that she developed her particular interest in special educational needs, co-authoring the Expanding Geography Scheme. Since 2001 Diane has worked on a variety of continuing professional development

initiatives for the Geographical Association as well as undertaking some PGCE tutoring and web-based advisory work for the Open University.

Jackie Wadlow

Jackie is currently headteacher and managing director of Alderwood Living and Learning with Autism Ltd. She is an Autism adviser and consultant to local education authorities and social services departments. She has worked in a variety of roles both in mainstream and special school context. She has trained as both an OFSTED team inspector, and trainer for NPQH and TEACCH (Treatment and Education of Autistic-related Communication Handicapped Children).

Contents of the CD

The CD contains activities and record sheets which can be amended/individualised and printed out by the purchasing individual or organisation. Increasing font size and spacing will improve accessibility for some students, as will changes in background colour. Alternatively, print onto pastel-coloured paper for greater ease of reading.

Introduction

Many teachers of geography feel daunted by the challenge of including students with special needs in their classrooms. This book is designed to make that challenge enjoyable and professionally rewarding.

> All children have the right to a good education and the opportunity to fulfil their potential. All teachers should expect to teach children with special educational needs (SEN) and all schools should play their part in educating children from the local community, whatever their background or ability. (*Removing Barriers to Achievement: The Government's Strategy for SEN, Feb. 2004*[1])

Geography has an essential contribution to make to the education of all young people. Its emphasis on place and on people ensures relevance through the use of a variety of teaching and learning styles. Geography has much to offer that is of value to all. As David Bell, Chief HMI (2005)[2] stated:

> Listen to a news broadcast or open a newspaper and you cannot fail to be struck by the relevance of geography. This practical discipline enables us to understand change, conflict and key issues which impact on our lives today and which will affect our futures tomorrow . . . If the aspiration of schools is to create pupils who are active and well-rounded citizens there is no more relevant subject than geography.

Recent legislation and statutory guidance has sought to make our mainstream education system more inclusive and ensure that students with a diverse range of ability and need are well catered for. This means that all staff involved in developing the geography curriculum need to have an awareness of how students learn and develop in different ways. They need to use this understanding to devise appropriate strategies to reduce and even remove barriers to achievement.

These barriers can often be overcome by the use of appropriate teaching styles, accessible teaching materials or thoughtful grouping of students. To support this an increasing awareness and understanding of an individual child's physical, sensory or cognitive impairments is essential. It is this developing understanding that is now shaping the legislative and advisory landscape of our education system, and exhorting all teachers to carefully consider their curriculum planning and classroom practice.

The revised regulations for SEN provision make it clear that mainstream schools are expected to provide for students with a wide diversity of needs, and teaching is evaluated on the extent to which all students are engaged and enabled to achieve. This book has been produced in response to the implications of all of this for secondary geography teachers. It has been written by a subject specialist with support from colleagues who have expertise within the SEN field so that the information and guidance given is both subject specific and pedagogically sound. The book and accompanying CD (see chapter 2 resources) provide a resource that can be used with colleagues:

- to shape departmental policy and practice for special needs provision
- enable staff to react with a measured response when inclusion issues arise
- to raise awareness as to how to break down the barriers to learning in geography

The major statutory requirements and non-statutory guidance are summarised in Chapter 1, setting the context for this resource and providing useful starting points for departmental INSET.

It is clear that provision for students with special educational needs (SEN) is not the sole responsibility of the Special Educational Needs Co-ordinator (SENCO) and his/her team of assistants. If, in the past, subject teachers have 'taken a back seat' in the planning and delivery of a suitable curriculum for these students and expected the Learning Support department to bridge the gap between what was on offer in the classroom and what they actually needed – they can no longer do so.

> All teaching and non-teaching staff should be involved in the development of the school's SEN policy and be fully aware of the school's procedure for identifying, assessing and making provision for pupils with SEN. (*Table of Roles and Responsibilities, Code of Practice, 2002*[3]).

Chapter 2 looks at departmental policy for SEN provision and provides useful audit material for reviewing and developing current practice. The term 'special educational needs' is now widely used and has become something of a catch-all descriptor – rendering it less than useful in many cases. Before the Warnock Report (1978[4]) and subsequent introduction of the term 'special educational needs', any students who for whatever reason (cognitive difficulties, emotional and behavioural difficulties, speech and language disorders) progressed more slowly than the 'norm' were designated 'remedials' and grouped together in the bottom sets, without the benefit, in many cases, of specialist subject teachers.

But the SEN tag was also applied to students in special schools who had more significant needs, and who had previously been identified as 'disabled' or even 'ineducable'. Add to these students with impaired hearing or vision, others with mobility problems, and even students from other countries with a limited understanding of the English language – who may or may not have been highly intelligent – and you have a recipe for confusion to say the least.

The day-to-day descriptors used in the staffroom are gradually being moderated and refined as greater knowledge and awareness of special needs is built up. The whole process of applying labels is fraught with danger. Sharing a common vocabulary – and more importantly, a common understanding – can help colleagues to express their concerns about a student and address these issues both in the geography classroom and during fieldwork activities. Often, this is better achieved by identifying the particular areas of difficulty experienced by the student, rather than by identifying the syndrome. The Code of Practice identifies four main areas of difficulty and these are detailed in Chapter 3. This chapter also includes an 'at a glance' guide to a wide range of syndromes and conditions and guidance on how they might present barriers to learning. These are accompanied by descriptions of real life situations with real students. This is a powerful way to demonstrate ideas and guidance. This chapter stresses the individuality of the learner.

There is no doubt that the number of students with special needs being educated in mainstream schools is growing:

> . . . because of the increased emphasis on the inclusion of children with SEN in mainstream schools the number of these children is increasing, as are the severity and variety of their SEN. Children with a far wider range of learning difficulties and variety of medical conditions, as well as sensory difficulties and physical disabilities, are now attending mainstream classes. The implication of this is that mainstream school teachers need to expand their knowledge and skills with regard to the needs of children with SEN. (Stakes and Hornby 2000[5])

The continuing move to greater inclusion means that all teachers can now expect to teach students with varied, and quite significant, special educational needs at some time.

Chapter 4 considers the components of an inclusive classroom and how the physical environment and resources, structure of the lesson and teaching approaches can make a real difference to students with special needs. This theme is extended in Chapter 5 to look more closely at teaching and learning styles and to consider strategies to help all students maximize their potential in outdoor learning. The requirement for students to engage in fieldwork is a statutory part of the National Curriculum key stages 1 to 4. It is also a way to provide different and appropriate stimuli to learning in geography, often aiding motivation. It provides many opportunities to work appropriately and flexibly with all students, including those with identified special needs.

The monitoring of students' achievements and progress is a key factor in identifying and meeting their learning needs. Those students who make slower progress than their peers are often working just as hard, or even harder, but their efforts can go unrewarded. Chapter 6 addresses the importance of target setting and subsequent assessment and review in acknowledging students' achievements and in showing the department's effectiveness in value-added terms.

Liaising with the SENCO and support staff is an important part of every teacher's role. The SENCO's status in a secondary school often means that this teacher is part of the leadership team and influential in shaping whole-school policy and practice. Specific duties might include:

- ensuring liaison with parents and other professionals
- advising and supporting teaching and support staff

- ensuring that appropriate Individual Education Plans are in place
- ensuring that relevant background information about individual students with special educational needs is collected, recorded and updated
- making plans for future support and setting targets for improvement
- monitoring and reviewing action taken

The SENCO has invariably undergone training in different aspects of special needs provision and has much to offer colleagues in terms of in-house training and advice about appropriate materials to use with students. The presence of the SENCO at the occasional departmental meeting can be very effective in developing teachers' skills in relation to meeting SEN, making them aware of new initiatives and methodology and sharing information about individual students.

In most schools, however, the SENCO's skills and knowledge are channelled to the chalkface via a team of Teaching or Learning Support Assistants (TAs, LSAs). These assistants can be very able and well-qualified, and deserve to be used purposefully in the classroom. Chapter 7 looks at how teachers can manage in-class support in a way that makes the best use of a valuable resource.

There is no doubt that the number of students with identified special educational needs being educated in mainstream schools is growing. The reaction of many staff, when faced with this reality, is to feel a lack of confidence. We hope that this text will help to minimise their understandable concerns.

References

1. Removing Barriers to Achievement: the Government's Strategy for SEN, February, 2004
2. Bell, D (2005) 'The value and importance of geography' *Teaching Geography,* 30, 1: 12–13
3. Table of Roles and Responsibilities, Code of Practice, 2002
4. Warnock Report. Department of Education and Science (1978) *Special Educational Needs: Report of the Committee of Inquiry on the Education of Handicapped Children and Young People.* London: HMSO
5. Stakes, R and Hornby, G (2000) *Meeting Special Needs in Mainstream Schools.* London: David Fulton Publishers.

CHAPTER 1

Meeting Special Educational Needs – A Responsibility for All

Introduction

> The human landscape can be read as a landscape of exclusion . . .. The simple questions we should be asking are: who are places for?, whom do they exclude?, and how are these prohibitions maintained in practice? (Sibley 1995[1])

Geography has an essential educational contribution to make for all. Lambert and Morgan (2005) state that

> When any teacher witnesses the penny drop in a student's mind or the focused buzz of 30 teenagers grappling with a mystery, it is very satisfying, but in geography (we are biased) it is doubly so – because often we are in the business of helping students make sense of the world as it is, to see it in new ways and gain the confidence to believe they could even change it. Learning geography, therefore, is a fine vehicle for education in a world in which issues of citizenship and sustainable development will gain a greater prominence during students' lifetimes[2].

It is therefore highly significant that careful thought should be given to how students with special educational needs are best supported in their learning of geography. We aim to offer help and advice that is of value to all staff connected with a geography department so that they can develop their curriculum in an appropriate way. We are acutely aware that time is not a commodity that many teachers of geography have to spare. However, increased understanding as to how small changes, such as altering the colours on a map, can improve the geographical experiences of some students, reaps rewards well beyond the cost of those few extra minutes. More than that we are convinced that some of the curriculum changes that result from the inclusion agenda will improve learning experiences for all.

The power and relevance of geography

An informed citizen must have an understanding of how the world works. This requires appropriate engagement with economic, social, political as well as physical and environmental processes. The power and relevance of an effective geographical education is illustrated below. This document has been constructed by the Geographical Association and has much to offer to all schools when considering the power and relevance of geography for all students, including those with identified special needs, and is on the accompanying CD.

THE POWER AND RELEVANCE OF GEOGRAPHY IN EDUCATION

The learning of geography is concerned with:

The physical world: land, water, air and ecological systems and the processes that bring about change in them. Can involve spiritual dimensions.

Human environments: societies and communities, and the human processes involved in understanding work, home, consumption and leisure. Involves political, moral and ethical dimensions.

Interdependence: spatial manifestations of interaction such as trade, migration, climate change: Involves, crucially, linking the 'physical' and 'human' and the emerging concept of 'sustainable development'.

Place and space: the 'vocabulary' and the 'syntax' of the world, developing knowledge and understanding of location and interconnectedness.

Scale: the lens through which the subject matter is 'seen'. Emphasises the significance of local, regional, national, international and global perspectives.

Students' lives: using students' images, experiences, meanings and questions can introduce an explicit futures orientation into lessons and 'reach out' to students as active agents in their learning.

David Lambert, Geographical Association, 2004

Additionally QCA[3] clarifies the importance of geography to students with learning difficulties.

> Learning geography helps students develop curiosity in, and an understanding of, themselves, other people and places, and the relationships between them. In particular, studying geography offers students with learning difficulties opportunities to:
>
> - become aware of, and understand their personal position in, space
> - become aware of, and interested in, themselves and their immediate surroundings

- explore local and then wider environments
- develop an interest in, and knowledge of, places and people beyond their immediate experience
- experience aspects of other countries and cultures, especially where there are comparisons with their own.

In response to these opportunities, students can make progress in geography by:

- increasing the breadth and depth of their experience and knowledge
- studying smaller (local scale) to larger areas (regional and national scales)
- extending studying from the familiar to the less familiar, for example, from their own locality to places which are further away
- gaining understanding, for example from understanding abstract as well as concrete concepts.

Geography well taught has a pivotal role to play for students with special educational needs. There are of course many challenges to be faced. Careful consideration must be made of how and what to teach. Much of this text is rightly concerned with how to present geographical information and perspectives. The content of geography, the discipline itself, is also significant. This is touched upon in several places. Geography is often presented from the perspective of the able. The geography curriculum offered in many schools would benefit from an audit, to explore how and when the geographies of the disabled are represented. A more inclusive representation of place would enable young citizens to become better informed. In this way, decisions that they make about places will be enriched by alternative perspectives. They will be able to consider how to improve places both physically and socially.

One way to achieve this is to engage all students in an accessibility mapping activity. This is described in more detail below. This activity is on the accompanying CD.

Making an Accessibility Map

Adapted from 'Disability, Space and Society', Rob Kitchin, Geographical Association, Sheffield, 2000[4].

Take a walk around a small part of your local shopping area. Find a good position to stand for 5 minutes:

- count the number of people you see go past you. How many appeared to be disabled?

Whilst walking consider:

- how people with physical and sensory impairment might get to this place
- how accessible the shops are to people with physical and sensory impairments
- are there any visual signs or colour codings that might help people to find their way around?
- is there a shop mobility system? Are there accessible toilets?

Now use this information to create an accessibility map of the area that you studied. Try to make your map as inclusive as possible so that it contains information for people with different disabilities. Remember that disabled people consist of more than just wheelchair users.

You will find a list of possible map symbols that could be used in this activity on the CD.

What does SEN mean?

The definition which is most commonly adopted is taken from Part 1 of the Disability Discrimination Act (DDA) 1995. Under the DDA, a person is considered disabled if they have 'a physical or mental impairment which has an effect on their normal day-to-day activities.' That effect must be:

- substantial (more than minor or trivial)
- long term (lasting for at least 12 months)
- adverse

'Physical or mental impairments' include:

- sensory impairments
- hidden impairments (e.g. dyslexia, learning difficulties, epilepsy, diabetes etc.)
- 'clinically well-recognised' mental illnesses

Geography and Special Needs

So what's the point of teaching and learning geography for students with special educational needs? To help us to answer this significant question carefully, it might be helpful to consider this description by Peter White[6] on his first journeys outside the gates of his special school for the blind. Peter White is now a respected media journalist, he is known in particular for his work for the BBC series on issues concerning disability, 'Does he take sugar?'

> In those early weeks my adventures included getting lost on almost every housing estate around the edges of Worcester, falling into several muddy ditches in the surrounding lanes . . . finding myself in a farmer's hen coop . . . In the end I shattered this happy state of bucolic mayhem . . . by getting run over by a rather large lorry . . . One minute I thought I was crossing the entrance to Dog Rose Lane in what was a permanent state of geographic uncertainty . . . In the end it turned out that all I had sustained was a bump on the head.

This extract supports us in thinking about all sorts of positive reasons for all students to access elements of a geographical education. Geography can help:

- students to develop a sense of their own space and place
- encourage a freedom of movement by developing a confidence about spaces and places
- develop an appropriate awareness of others, other people, other spaces
- develop a sense of independence within spaces and places and a sense of interdependence through an appropriate awareness of others
- develop a sense of awe and wonder about the world
- students to learn about places and their similarities and differences, appreciating that difference and diversity are an asset rather than a threat
- where appropriate to develop an awareness that each of us is a citizen of the world and that we have rights and responsibilities
- to establish links with others, via e-mails or other forms of exchange and visits

Below, Bev Rowley, from Stretton Brook School, shares the aims that she has developed for geography and special needs at her school. This is also on the accompanying CD.

Aims of geography at Stretton Brook School

Geography is a foundation subject in the National Curriculum and is therefore part of the entitlement of all our students between five and fourteen years of age. We believe it has an important part to play in the education of students with special needs. By studying places, and the human and physical processes that shape them, at a variety of scales students develop a sense of awe and wonder and improve their geographical understanding. By thinking about the people who live in and connect with these places our students are supported in making more sense (both through appropriately communicated geographical information and sensory experiences) of their personal space, immediate surroundings and the wider world. For this reason geography is also a subject studied by students in the 14–16 year age group. Currently many students come from backgrounds that offer limited insights into locations beyond their immediate home environment. Therefore, it is our aspiration to provide learning experiences and strategies that offset this.

Our aims are:

- To help develop geographical understanding through the provision of information communicated in an appropriate way.
- To introduce students to geographical enquiry and to arouse their interest and curiosity in their personal space, immediate locality and in the wider world.
- To help students develop a sense of identity through learning about their locality, home region and the UK and its relationship and interconnections with other countries.
- To foster in our students a better understanding of different cultures both within our own society and elsewhere in the world.
- To help students to understand how people use and misuse their environments.

These aims can be achieved by:

- Recognising, including, valuing and building on a student's own experience.
- Teaching about real places and the people who live in them by using place studies from a variety of locations and at a variety of scales.
- Encouraging students to be engaged in real questions about issues and problems relating to people in different places and how these interconnect with our own lives and our own places, i.e. use of key questions, an enquiry approach.
- Allowing opportunities for active learning so that students can see the purpose of their investigations and are then motivated to pursue them – use of fieldwork, local trails, CD-ROMs, internet, and other information sources.
- Enabling students to make connections with the wider curriculum and their learning outside of school.

Policy into practice

In many cases, students' individual needs will be met through greater differentiation of tasks and materials, i.e. school-based intervention as set out in the SEN Code of Practice (2001[7]). All schools should have a Special Educational Needs Co-ordinator (SENCO) and maintain a register of students with a special educational need. The Code of Practice outlines a staged approach to responding to students' needs. This consists of:

School Action where the school makes appropriate provision from its own expertise and resources. This may involve the class teacher identifying that a student needs extra support and then preparing a plan for the student with the school SENCO.

School Action Plus where the school makes appropriate provision with the support of external expertise e.g. specialist teacher, educational psychologist. The school will develop an individual education plan (IEP) with support from other professionals.

Statementing where a local education authority undertakes an assessment of a student's needs and, if appropriate, establishes a formal statement. This defines the student's needs and the provision to be made.

Julie Dale has developed a PowerPoint presentation that she uses to share this information at parents' evenings in her school, St. John Fisher High School, Newcastle under Lyme. This is available on the CD.

What does the Act cover?

Part 4 of the Special Educational Needs and Disability Act, 2001[8] was implemented from September 2002 and becomes Part 4 of the Disability Discrimination Act. The Act covers pre- and post-16 education.

The Act introduces the right for disabled students not to be discriminated against in education, training and any service provided wholly or mainly for students or those enrolled on a course when this is provided by certain 'responsible bodies' such as schools, colleges or LEAs. This means that the Act covers not only education, but other areas as well. These may include:

- arranging study abroad or work placements
- careers advice and training
- chaplaincy and prayer areas
- learning equipment and materials such as laboratory equipment, computer facilities, class handouts etc.
- outings and trips
- libraries, learning centres and information centres and their resources
- examinations and assessments
- field trips
- informal/optional study skills sessions
- information and communication technology and resources

From 1 September 2003, responsible bodies are also required to make adjustments that involve the provision of auxillary aids and services, and after 1 September 2005 responsible bodies are required to make adjustments to the physical features of premises where these put disabled people or students at a substantial disadvantage. Of course, some of the services covered by this new Act were previously covered by part III of the DDA 1995 but will now come under the new Act as part IV of the DDA and will be more strictly enforced.

The duty to make 'reasonable adjustments' is a duty to disabled people generally and not just to particular individuals. This will be of benefit to society as a whole. This 'anticipatory' aspect effectively means that providers must consider what sort of adjustments may be necessary for disabled people in future and, where appropriate, make adjustments in advance. These might include:

- changes to policies and practices
- changes to course requirements
- changes to physical features of a building
- provision of interpreters or other support workers
- the delivery of courses in alternative ways
- the provision of materials in other formats

Anticipatory action for a geography department could include producing handouts and course materials in an electronic format in order that they can be adapted in a variety of ways. This might involve enlarging the font, printing on different coloured paper, simplifying diagrams. It also enables appropriate pictorial symbols to be included.

An example of a reasonable adjustment to an activity is described here. As part of a geography course students are required to stay overnight in a basic field study centre. A student who needs regular dialysis cannot go on the field trip without having her treatment. A reasonable adjustment might be for the teacher to arrange for her to take part during the days but for a responsible and suitably qualified adult to return with her to a nearby town or village in order that she can have her dialysis each evening.

The inclusive school and department

The Index for Inclusion[9] asks us to consider how we work and how we teach. 'Inclusion in education involves the process of increasing the participation of students in, and reducing their exclusion from, the cultures, curricula and communities of local schools.'

The basis of the Index is to create a new language whereby the concept of 'special educational needs' is replaced by the term 'barriers to learning and participation'. In real terms, this represents a shift from a medical model of difficulties in education to a social one. It is about students' strengths and their potential contribution. It is not about a deficit model. This is to be welcomed. Schools, they argue, can have an impact on reducing disabilities due to physical, personal and institutional barriers to access and participation.

This resonates with the Key Stage 3 strategy[10] (this will be known as the Secondary Strategy from September 2005). 'All schools will want to attach particular importance to promoting mutual respect and understanding of different religions, cultural traditions and languages.'

An example of such work can be found through the Geographical Association's Valuing Places project. Here affective mapping has been used to support students in valuing their environment. 'Affective mapping means plotting on maps the feelings that particular places evoke. Feelings are shown by symbols, possibly supplemented with annotation. Where patterns of feelings can be identified on maps, then areas can be shaded accordingly.'[11] Here Ruth Barton from the Vale of Evesham School[12] describes how she has adapted this activity.

> I teach a group of students with specific learning, behavioural and communication problems and decided to try out the idea of Affective Mapping with a Year 7 group. We linked hands and travelled around the school as a team. This brought a sense of unitedness and prevented any students becoming distracted or not on task.
>
> The areas of the school visited were then given a 'thumbs up', an 'OK' or a 'thumbs down' card. The thumbs up meant the area felt safe and good to be in. The OK one meant it was neither good nor bad, just average. The thumbs down card meant the environment felt unsafe or uncomfortable to be in.
>
> The results were as follows:

	Thumbs up	OK	Thumbs down
Classroom	✓ Vocabulary – Safe		
Corridor		✓ okay	
Toilets			✓ bullying
Entrance Foyer	✓ safe – calm		
Dining Room			✓ noisy
Playground		✓ okay, but noisy	
Car Park			✓ dangerous
Path	✓ safe		

> Our School Council representative is going to take this information to our half termly main meeting and with help from other Council members and Senior

Management Team address the issues, e.g. dining room too noisy – thumbs down.

Other classes could do the same exercises and feel empowered to change their school environment or use it for the better.

I really feel this is a very valuable exercise.

Summary

Students with a wide range of needs – physical, emotional, cognitive and social – are to be found in every classroom in England. Learning geography can be a challenge, but it is a significant factor in helping students to appreciate their place, different places, communities and cultures. Government policies only point the way. It is ultimately teachers who make the rhetoric into reality.

References

1. Sibley, D (1995) *Geographies of Exclusion: Society and difference in the West*, ix. London: Routledge
2. Morgan, J and Lambert, D (2005) *Teaching School Subjects: Geography.* London: RoutledgeFalmer
3. QCA (2001) *Planning, Teaching and Assessing the Curriculum for Pupils with Learning Difficulties*: Geography. London: QCA
4. Kitchin, R (2000) *Disability, Space and Society.* Sheffield: Geographical Association
5. Disability Discrimination Act (1995) London: HMSO
6. White, P (2000) *See It My Way.* Time Warner Paperbacks
7. Special Educational Needs Code of Practice (2001) London: DfES/581/2001
8. Special Educational Needs and Disability Act (2001) www.disability.gov.uk/policy/sen/
9. Index for Inclusion (2000) http://inclusion.uwe.ac.uk/csie/indexlaunch.htm
10. Key Stage 3 Strategy, www.standards.dfes.gov.uk/keystage3/
11. Roberts, M (2003) *Learning Through Enquiry.* Sheffield: Geographical Association
12. Barton, R (2005) 'Inclusion and emotional mapping' *Teaching Geography,* 30,1: 39

Additional references

www.geography.org.uk
www.geographyshop.org.uk
www.qca.org.uk/geography
www.hmso.gov.uk/dda/
www.disability.gov.uk
www.teachernet.gov.uk/_doc/3724/SenCodeOfPractice.pdf

CHAPTER 2

Departmental Policy – Underpinning Potential Actions

Introduction

Devising a policy that outlines strategies for meeting students' special educational needs in geography will have many benefits beyond the useful document that results. The dialogues involved in creating such a document will be of huge value. The policy, itself, should set the scene for any visitor to the geography department, from supply staff to inspectors, and make an informed contribution to the departmental handbook. The process of developing a departmental SEN policy offers the opportunity to clarify and evaluate current thinking and practice within the geography team and to establish a consistent approach.

The policy should:

- clarify the responsibilities of all staff and identify any with specialist training and/or knowledge
- describe the curriculum on offer and how it can be differentiated
- outline arrangements for assessment and reporting
- guide staff on how to work effectively with support staff
- identify staff training

The starting point will be the school's SEN policy as required by the Education Act 1996[1], with each subject department 'fleshing out' the detail in a way which describes how things work in practice. The writing of a policy should be much more than a paper exercise completed to satisfy the senior management team and OFSTED inspectors: it is an opportunity for staff to come together as a team and create a framework for teaching geography in a way that makes it accessible to all students in the school (see Appendix 2.1, INSET activity, also on the CD).

An extract from the Special Educational Needs Policy developed by St. John Fisher's Geography Department follows on page 16.

St. John Fisher's Geography Department Policy on Special Educational Needs

The department's vision is to give all students full access to the Geography National Curriculum to enable them to achieve to their full potential.

This will be achieved by:

- Identifying the needs of the students at the earliest opportunity, working within the school's SEN policy on identification and assessment.
- Ensuring that the needs of students are made known to all geography staff. This is facilitated by using the information produced by the SEN department, namely the strategies booklet.
- Giving consideration to the appropriate resources to support students, staff and the curriculum. Developing worksheets and expanding resources for low ability and gifted students.
- Review student progress regularly, reporting concerns to the Head of Department.
- Appropriate assessment recording and reporting so that students feel valued and are given constructive feedback on how to improve.
- Appropriate consultation with parents.
- Liaison with Heads of Year and Teaching Assistants.
- Encouraging relevant INSET for each member of the department.
- Ensuring SEN students join in all the general activities of the department.
- Departmental representation at SEN link meetings.
- Teaching students in ability sets and setting students where possible in Key Stage 4 to allow the style of teaching to be adapted to the ability of the group.
- Involving Teaching Assistants in fieldwork where appropriate.
- Acting on IEP information and applying this to the teaching methodology.

This document is supported by a 'Special Needs Link File'.
This file contains:

- SEN Policies, including referral and praise forms [available on the CD and in the appendix].

- Information and extracts to support the development of literacy and numeracy in geography.
- SEN Resource audit, including copies of booklets and resources suggested.
- Strategies booklet, including a variety of teaching and learning ideas.
- Current SEN register.
- Minutes of SEN Link representative meetings.

This provides us with useful information as to one school's approach; the rest of this chapter will support you in constructing your own policy. The next section provides you with some initial ideas.

Where to start when writing a policy

An audit can act as a starting point for reviewing current policy on SEN or to inform the writing of a new policy. It will involve gathering information and reviewing current practice with regard to students with SEN and is best completed by the whole of the department, preferably with some additional advice from the SENCO or another member of staff with responsibility for SEN within the school. An audit carried out by the whole department can provide a valuable opportunity for professional development, if it is seen as an exercise in sharing good practice and encouraging joint planning. But before embarking on an audit, it is worth investing some time in a department meeting or training day, to raise awareness of special educational needs legislation and establish a shared philosophy. The accompanying CD and appendix contain an activity to use with staff.

The following headings may be useful in establishing a working policy:

General statement

- What does legislation and DFES guidance say?
- What does the school policy state?
- What do members of the department have to do to comply with it?

Definition of SEN

- What does SEN mean?
- What are the areas of need and the categories used in the Code of Practice?
- Are there any special implications within the subject area?

Provision for staff within the department

- How is information shared?
- How is the shared responsibility for SEN negotiated?

- Who has responsibility for leading SEN?
- How and when is information shared?
- Where and what information is stored?

Provision for students with SEN

- How are students with SEN assessed and monitored in the department?
- How are contributions to IEPs and reviews made?
- What criteria are used for organising teaching groups?
- What alternative courses are offered to students with SEN?
- What alternative resources are offered to students with SEN?
- What special internal and external examination arrangements are made?
- What guidance is available for working with support staff?

Resources and learning materials

- Is there any specialist equipment used in the department?
- How are resources developed?
- Where are resources stored?

Staff qualifications and Continuing Professional Development needs

- What qualifications do the members of the department have?
- What training has taken place?
- How is training planned?
- Is a record kept of training completed and training needs?

Monitoring and reviewing the policy

- How will the policy be monitored?
- When will the policy be reviewed?

The content of a SEN departmental policy

This section gives detailed information on what a SEN policy might include. Each heading is expanded with some detailed information and raises the main issues with regard to teaching students with SEN. At the end of each section there is an example statement. The example statements can be personalised and brought together to make a policy.

General statement with reference to the school's SEN policy

All schools must have a SEN policy according to the Education Act 1996. This policy will set out basic information on the school's SEN provision, and how the school identifies, assesses and provides for students with SEN, including information on staffing partnerships with other professionals and parents.

Any department policy needs to have reference to the school SEN policy.

Example

All members of the department will ensure that the needs of all students with SEN are met, according to the aims of the school and its SEN policy.

Definition of SEN

It is useful to insert at least the four areas of SEN in the department policy, as used in the Code of Practice for Special Educational Needs[2].

THE FOUR AREAS OF SEN

Cognition and Learning Needs	Behavioural, Emotional and Social Development Needs	Communication and Interaction Needs	Sensory and/or Physical Needs
Specific Learning Difficulties (SpLD) Dyslexia Dyscalculia Dyspraxia Moderate Learning Difficulties (MLD) Severe Learning Difficulties (SLD) Profound and Multiple Learning Difficulties (PMLD)	Behavioural, Emotional and Social Difficulties (BESD) Attention Deficit Disorder (ADD) Attention Deficit Hyperactivity Disorder (ADHD)	Speech, Language and Communication Needs Autistic Spectrum Disorder (ASD) Asperger's Syndrome	Hearing Impairment (HI) Visual Impairment (VI) Multi-Sensory Impairment (MSI) Physical Difficulties (PD)

Provision for staff within the department

In many schools, each department nominates a member of staff to have special responsibility for SEN provision (with or without remuneration). This can be very effective where there is a system of regular liaison between department SEN representatives and the SENCO in the form of meetings or paper communications or a mixture of both.

The responsibilities of this post may include liaison between the department and the SENCO, attending any liaison meetings and providing feedback via meetings and minutes, attending training, maintaining the departmental SEN information and records, and representing the need of students with SEN at departmental level. This post can be seen as a valuable development opportunity for staff. The name of this person should be included in the policy.

How members of the department raise concerns about students with SEN can be included in this section. Concerns may be raised at specified departmental meetings before referral to the SENCO. An identified member of the department could make referrals to the SENCO and keep a record of this information.

Reference to working with support staff will include a commitment to planning and communication between staff. There may be information on inviting support staff to meetings, resources and lesson plans.

A reference to the centrally held lists of students with SEN and other relevant information will also be included in this section. A note about confidentiality of information should be included.

Example

> The member of staff with responsibility for overseeing the provision of SEN within the department will attend liaison meetings and feedback to other members of the department. Other responsibilities will include maintaining the department's SEN information file, attending appropriate training and disseminating this to all departmental staff. All information will be treated with confidentiality.

Provision for students with SEN

It is the responsibility of all staff to know which students have SEN and to identify any students having difficulties. Students with SEN may be identified by staff within the department in a variety of ways, these may be listed and could include:

- observation in lessons
- observations whilst working outside of the classroom
- assessment of class work
- homework tasks
- end of module tests
- progress checks

- annual examinations
- reports

Setting out how students with SEN are grouped within the geography department may include specifying the criteria used and/or the philosophy behind the method of grouping.

Example

> The students are grouped according to ability as informed by Key Stage 2 results, reading scores and any other relevant performance, social or medical information.

Monitoring arrangements and details of how students can move between groups should also be set out. Information collected may include:

- National Curriculum levels
- departmental assessments
- reading scores
- advice from pastoral staff
- discussion with staff in the SEN dept
- information provided on IEPs

Special Examination arrangements need to be considered not only at Key Stages 3 and 4 but also for internal examinations. How and when these will be discussed should be clarified. Reference to the SENCO and examination arrangements from the examination board should be taken into account. Ensuring that staff in the department understand the current legislation and guidance from central government is important, so a reference to the SEN Code of Practice and the levels of SEN intervention is helpful within the policy. Here is also a good place to put a statement about the school behaviour policy and rewards and sanctions, and how the department will make any necessary adjustments to meet the needs of students with SEN.

Example

> It is understood that students with SEN may receive additional support if they have a statement of SEN, are at School Action Plus or School Action. The staff in the Geography department will aim to support the students to

achieve their targets as specified on their IEPs and will provide feedback for IEP or Statement reviews. Students with SEN will be included in the departmental monitoring system used for all students. Additional support will be requested as appropriate.

Resources and learning materials

The department policy needs to specify what differentiated materials are available, where they are kept and how to find new resources. This section could include a statement about working with support staff to develop resources or access specialist resources as needed, and the use of ICT. Teaching strategies may also be identified if appropriate. Advice on more specialist equipment can be sought as necessary, possibly through LEA support services: contact details may be available from the SENCO, or the department may have direct links. Of course, ICT plays a vital role in enhancing learning and teaching in geography. Both the internet and CD-ROMs are used to investigate places and topics. These should be assessed for their accessibility by different students. Databases are used both to investigate and present geographical information. E-mail can be used to communicate with people in other localities. This should be carefully supported by staff. Additionally, any specially bought subject text or alternative/appropriate courses can be specified as well as any external assessment and examination courses. On the CD is an example as to how differentiation has been expressed by Stretton Brook's geography department.

Example

The department will provide suitably differentiated materials and, where appropriate, specialist resources for students with SEN. Additional texts are available for those students working below National Curriculum level 3. At Key Stage 4, an alternative course to GCSE is offered at Entry level but, where possible, students with SEN will be encouraged to reach their full potential and follow a GCSE course. Support staff will be provided with curriculum information in advance of lessons and will also be involved in lesson planning. A list of resources is available in the department handbook and on the notice board.

Staff qualifications and Continuing Professional Development needs

It is important to recognise and record the qualifications and special skills gained by staff within the department. Training can include not only external courses but also in-house INSET and opportunities such as observing other staff, working to produce materials with other staff, and visiting other establishments. There are also subject specialist networking opportunities through the Geographical Association. Staff may have hidden skills that might enhance the work of the department and the school: for example, some staff might be proficient in the use of sign language, some TAs may have undertaken a specialist qualification in dyslexia, whilst others may have particular strengths regarding fieldwork and out of classroom learning.

Example

A record of training undertaken, specialist skills and training required will be kept in the department handbook. Requests for training will be considered in line with the department and school improvement plan.

Monitoring and reviewing the policy

Any policy to be effective needs regular monitoring and review. These can be planned as part of the yearly cycle. The responsibility for the monitoring can rest with the Head of Department, but will have more effect if supported by someone from outside acting as a critical friend; this could be the SENCO or a member of the senior management team in school.

Example

The Department SEN policy will be monitored by the Head of Department on a planned annual basis, with advice being sought from the SENCO as part of a three-yearly review process.

Summary

Creating a departmental SEN policy should be a developmental activity to improve the teaching and learning for all students but especially for those with special or additional needs. The policy should be a working document that will evolve and change; it is there to challenge current practice and to encourage improvement for both students and staff. If departmental staff work together to create the policy, they will have ownership of it; it will have true meaning and be effective in clarifying practice.

References

1. Education Act (1996) www.hmso.gov.uk/acts/acts1996/1996056.htm
2. Special Educational Needs Code of Practice (2001). London: DfES/581/2001

CHAPTER 3

Different Types of SEN

Introduction

This chapter aims to help staff in mainstream schools to identify the basic characteristics of a range of disabilities. It will attempt to highlight patterns of learning and suggest some teaching strategies that enhance access and support for students with special needs. This section is not intended to provide staff with all the answers or turn them into specialists. This would, in fact, be an impossible task, a task that becomes even more complex with our acknowledgement that all students are unique; and whilst two students may have the same diagnosis the barriers to learning, or the support they need, may differ. It is always an advantage to seek expert advice whenever possible. Staff should establish what internal and external support is available and what local or national societies can offer, e.g. SENCO, LEA Advisors, National Society for Dyslexia etc. However, some key knowledge of the impact of the disability is essential to avoid unnecessary hardship or difficulties for both students and staff.

The SEN Code of Practice (DfES 2001[1]) suggests that it is helpful to see students' needs as being in four broad groups, with further sub-divisions based on the categories used by Ofsted as follows (see also page 19 in Chapter 2):

1 Cognition and Learning

2 Behavioural, Emotional and Social Difficulties

3 Communication and Interaction

4 Sensory and/or Physical Impairment.

In reality some students may present characteristics that would fit into one or more of these categories or indeed all. In addition a diagnosis that relates to the majority of the categories within the framework is not an indicator of the severity of any specific condition, or of a student's academic ability which may cover the whole continuum. All students are individuals, and all will have some days that are better than others. The recency of their diagnosis, their particular strengths and weaknesses, their course choices (Key Stage 4 and beyond) and coping strategies will all have an impact.

There are of course other students with what may be termed 'hidden disabilities' such as seasonally affected medical conditions, including allergies, glue ear and asthma. The geography teacher needs to be aware of these, as they may affect learning and performance, and may require intermittent adaptations to teaching or resources, especially when considering field studies.

Many of the aspects of good practice already implemented within mainstream settings to meet the needs of all students will also be effective in supporting students with special needs. Some of these have been listed below:

- Inclusive school ethos that celebrates diversity.
- Effective general and specific training for all staff and additional specialised support when and if required.
- Multi-agency support, the involvement of professionals from several different areas of expertise, including education, health, social services etc.
- A geography curriculum that provides access to students of all abilities including SEN.
- Differentiation at all levels of planning that meets the needs of students of all abilities.
- Planning of lessons that have the same content, but where the style of teaching is adapted to the abilities of the class. Differentiation is gained through both input and output.
- Appropriate resources to support students, staff and the curriculum.
- Use of stimulating resources and contexts which engage the learner, support conceptual development and meaning, and give a real sense of place through first hand experiences.
- Effective assessment systems that identify students' needs at the earliest opportunity, meet the wide range of students' needs and identify progress and achievement, and provide on-going monitoring and support to students and staff.
- Flexible teaching styles and strategies.
- The development of effective teaching teams which, where appropriate, includes and values the role of support staff.
- Discussions with students to identify their particular access difficulties and adapt teaching and learning styles and resources appropriately.
- Working closely with parents.
- Involvement of students within their own target setting.

- Where appropriate, the use of IEP information and applying this to the teaching methodology.

We are convinced that appropriate geography activities are essential to the development of all students. We are acutely aware that such experiences are not always easy to construct. Appendix 3 could be used at a departmental meeting or introduced during a meeting with the SENCO or whilst liaising with teaching assistants. It could be used to stimulate dialogue and informed discussion about the opportunities and constraints of a geographical education for students with special needs.

With more complex special needs additional and specific strategies may be required to break down barriers to learning. It is always valuable to develop a student profile in order to enable the teacher not to lose sight of the impact of the disability, and the fact that these are individual students. An example is given below. We have used this profile to organise our thinking about the different types of special educational need and the role that the geography teacher can play in reducing the barriers to learning. Each profile including this sample appears on the CD in order that you can (if you feel that it is helpful to your own teaching and learning circumstances) use these to create your own documents.

<table>
<tr><td>Name</td><td>D.O.B</td><td>Year Group</td><td>Form Group</td><td>Diagnosis</td></tr>
<tr><td colspan="3">What are the characteristics?</td><td colspan="2"></td></tr>
<tr><td colspan="3">What is the impact of these characteristics on learning?</td><td colspan="2"></td></tr>
<tr><td colspan="3">Supportive Teaching and Learning Strategies</td><td colspan="2"></td></tr>
<tr><td colspan="3">Preparation and planning for outdoor learning</td><td colspan="2"></td></tr>
</table>

There may well of course be some students for whom the mainstream environment will not fully meet their learning requirements and they may need a more specialist setting.

Within the profiles are some helpful hints as to how the geography teacher can best support access to the curriculum and learning for students who exhibit these particular characteristics.

These profiles, if used and adapted for individual students within your school, should of course become working documents. They will change as the student does.

The student perspectives have been constructed by the Valuing SEN team. These are based on conversations and observations with students that they teach. Of course names and some details have been altered, but the pen portraits are real reflections of learning needs.

SPECIFIC LEARNING DIFFICULTIES (SpLD)

The term 'specific learning difficulties' covers dyslexia, dyscalculia and dyspraxia.

Dyslexia (SpLD)

Dyslexia is a neurological disorder that disrupts the brain's ability to process language. It is a complex condition that can significantly and persistently affect a person's ability to cope with reading, writing or spelling. Dyslexia can occur at any level of intellectual ability and it is important to remember that the impact on learning will vary from individual to individual.

Main characteristics:

- The student may frequently lose their place while reading, make a lot of errors with the high frequency words, have difficulty reading names, and have difficulty blending sounds and segmenting words. Reading requires a great deal of effort and concentration.
- The student's work may seem messy with crossing outs, similarly shaped letters may be confused, such as b/d/p/q, m/w, n/u, and letters in words may be jumbled, such as tired/tried. Spelling difficulties often persist into adult life and these students may become reluctant writers without appropriate support.

Sam's Perspective on Dyslexia

I would love to read age-appropriate stories about places and discover further information about my particular interests in geography, rocks! I am pleased that I have found part of the answer to this problem through the use of listening books. I find it very difficult to read and understand a written passage but if someone else reads it to me and I follow the words I am able to answer the question much more quickly and waste less time. I hate trying to use punctuation, often get confused with upper and lower case letters and I reverse some letters. I am really bad at spelling and I usually learn how to spell a word by looking for patterns in the way the letters are arranged. This takes a long time and I always forget how to spell the word when I need to use it in written work. This is extremely frustrating as I am able to remember other facts. I use a computer and spell check that helps, but sometimes the word I have written is so different from the correct spelling that the wrong word is found.

My written work is often messy because I often have to cross things out when I get them wrong or lose my place. It annoys me when teachers judge me on my written work and reading and not on what I know. I get very tired by the end of the day and feel useless and I am not.

DYSLEXIA PROFILE	
Potential impacts on learning	• Often loses place when moving from one line to the next; print often becomes blurred or letters or words appear to move. • Finds it difficult to obtain meaning from text and misreads questions. • Finds it very difficult to work from the board. • Finds it difficult to listen for long periods of time. • Becomes stressed and frustrated because is unable to write legibly, quickly or neatly and the content does not relate to knowledge. • Has difficulty in remembering daily/weekly timetables, instructions, facts for tests and examinations. • Rarely completes work on time. • Suffers from eye discomfort and headaches and is often tired. • Is creative and imaginative.
Possible supportive teaching and learning strategies	• When possible use a multi-sensory approach, draw on supplementary visual, tactile, and auditory resources in addition to text. For example the use of artefacts found in different environments, use of music and sounds relating to the places and themes being studied. • It is often helpful if the first words of each sentence are highlighted in colour and work is left on board for longer periods of time. If the geography department has electronic copies of its handouts, then worksheets can easily be altered; the use of PowerPoint on an interactive whiteboard can also be helpful here. • If place-based websites are being used, it may be worth exploring if the text colours on these can be altered. • Students with dyslexia may prefer to use oral ways to communicate what they have learnt. • Use of handouts in addition to whiteboard (on which it is best to use coloured pens, try blue rather than black) that are differentiated in terms of readability, levels of text, number of sentences on each page, font type and size, e.g. increase in spacing between words and lines, colour of paper, e.g. yellow or blue. • The use of recycled paper can create a geographical discussion as well as reduce glare. • Remember also to review the text used on maps and diagrams and other forms of visual presentation used in geography. When a handout is text-rich consider limiting the amount of information and presenting it in a vertical rather than linear format. • Text pre-read by member of staff who is familiar with the student's specific needs.

	• Use of mind mapping techniques may be helpful. • Teaching of specific strategies may be helpful, including alternative methods of recording, i.e. note taking, tape recorder, ICT, reading for meaning, revision skills. This is worth particular consideration when preparing any work to be done on field trips or outside of the usual classroom environment. • Keep a geography dictionary to support an understanding of new words and their meaning. • The use of spreadsheets can provide a visual aid to students with dyslexia as they can be taught visual methods of laying out their work. • Only ask the student to read aloud if you are certain that it is a manageable task. This will need careful consideration when engaged in some thinking through geography activities such as the use of mysteries and card sorting activities for cause and effect. • High expectations and low stress atmosphere. It is helpful (to all) to give a framework to the lesson. Be specific and try to link details and examples back to the overall concept; for example, when considering coasts and erosion, link place studies such as Studland in Dorset to some of the big ideas, for example, erosion and the notion of a dynamic earth.
Preparation and planning for outdoor learning	The above list is also helpful when considering access to fieldwork. Additionally consider: • The physical location, weather conditions (bright sunlight) when students are being asked to record observations or to collect data. • The amount of time allowed to complete tasks – is it sufficient to enable students to access learning? • The use of ICT for drafting, editing and publishing. The use of digital images and video relating to the new fieldwork locations can provide helpful support, facilitating access outside of normal lesson time to new and unfamiliar places. • A dyslexic student, when asked about their fieldwork experiences, stated 'Fieldwork is the best thing about geography, a good way to learn, but it's difficult for me to make notes in the field. Please put more information in the handouts and remember that it is hard for me to read and write in bright sunlight.'

The British Dyslexia Association Tel: 0118 966 8271
Website: www.bda-dyslexia.org.uk
Dyslexia Institute Tel: 01784 222 300 Website: www.dyslexia-inst.org.uk

Dyscalculia (SpLD)

There are a number of views relating to whether dyscalculia and dyslexia are interrelated or discrete specific learning difficulties. As it is beyond the scope of this chapter to debate these issues, the information below identifies some basic characteristic of dyscalculia and how they may impact on teaching and learning in geography.

The term 'dyscalculia' refers specifically to difficulties in acquiring mathematical skills and developing number concepts, facts and procedures.

Main characteristics:

- In numeracy, the student may have difficulty counting by rote, writing or reading numbers, miss out or reverse numbers, have difficulty with mental maths, and be unable to remember concepts, rules and formulae.
- In maths-based concepts, the student may have difficulty with money, telling the time, with directions, and with right and left, which are all key abilities to analysing data in the field or outside of the classroom.

Shajid

My name is Shajid and I am a 15-year-old boy with dyscalculia. Many people have not heard about this but it means that I have a mathematical learning disability and only when you have this problem do you realise that numbers are everywhere. With anything involving numbers in geography, even simple adding and subtracting, I become very confused and make a great many mistakes. For example I have a great problem with 'o's' and if I am working with a number such as 301 I can see it in my mind but if I need to write it down it becomes all mixed up, e.g. 310 or 130. If I am given geographical data verbally I am unable to picture the numbers in my head. If it is written down at least I have a chance but sometimes even though I know how to solve a problem, or have a diagram to help me through the small steps of a task, I may miscopy the numbers, or place the answer next to the wrong number. I always check and double-check my answers but I am unable to see that I have used the wrong sequence or term. Numbers are not the only things that I find difficult; reading the time on ordinary clocks, drawing shapes, following directions and identifying left and right remain a mystery. However, I have found some things that are helpful. These include using a digital watch and always wearing it on my left hand; this helps me with directions. I find plans and templates that help me know what to do next useful. Teachers who support me in my use of a calculator and computer also help to improve my learning.

I am lucky because I do not also have problems with words like some of my friends with dyscalculia. I am also very good at practical things and as my teachers say, I work hard, do my very best and have a great sense of humour. Working with my parents, teachers and friends I enjoy the challenge of finding different ways of beating my dyscalculia.

DYSCALCULIA PROFILE	
Potential impact of characteristics on learning	When using mathematical concepts in relation to geography, students may find it difficult to: • Remember rules, formulae, order of calculations, and basic addition, subtraction, multiplication and division facts. • Read, write and record in mathematics figures and facts without making mistakes. • Understand the technical language of mathematics even though they may understand the words in other contexts. • Remember the 'layout' of things and geographical locations. • Demonstrate a sense of direction and may easily become disorientated in new situations.
Possible supportive teaching and learning strategies	• Whenever possible use concrete rather than abstract mathematical examples to illustrate a problem, for example use pictures/models when dealing with traffic data. • Prepare a prompt sheet electronically, with worked examples of regularly used mathematical tools that are used and applied in geography. Students can refer to this at any time. • Work through problems with the student, first breaking them down into small steps and allowing time for checking. This is particularly significant for students when they come to undertake their GCSE coursework. Ask for a TA to work alongside students during this time and carefully consider how to enable peer support both whilst gathering data in the field and analysing it back in school. Seek advice from the staff and TAs who work with the Maths department, so that consistent support can be given. • Provide supplementary visual resources to support the mathematical processes or rules. Use pictures, photographs and models whenever possible. • When using maps and scale for decision-making, cardboard templates drawn to scale to be used with the map, e.g. when siting a new development, the proposed shopping centre is produced on a cardboard template to scale to support the student in locational decision-making. • Artefacts appropriate to the places and themes being studied provided whenever possible. • Support through the use of a calculator. Students given extra time to undertake coursework tasks. This will be explained thoughtfully to their peers so that they understand that this is fair and reasonable. • Rough paper always available for working out.

	• Extra practice given on data related tasks, including supplementary activities that involve counting objects rather than just dealing with numbers.
Preparation and planning for outdoor learning	• Review planning and tasks with students prior to the visit, identify possible issues and provide adapted material to address these. • Reinforcement for work on left/right, map work and directions provided prior to any fieldwork experience. • Students not left on their own in the field – a peer buddy system operates. This is to support them with their sense of direction. These students have their own strengths too, so careful thought needs to be given to the groups that they work with in order that they help each other.

www.dyscalculia.co.uk

Dyspraxia (SpLD)

Generally, 'dyspraxia' is defined as an immaturity of the organisation of movement which includes both fine and gross motor skills. Some students may also find difficulties with language and perception.

Main characteristics:

- difficulty in co-ordinating movements, may appear awkward and clumsy
- difficulty with handwriting and drawing, throwing and catching
- difficulty following sequential events, e.g. multiple instructions
- may misinterpret situations, take things literally
- limited social skills resulting in frustration and irritation
- some articulation difficulties (verbal dyspraxia)

A typical day for James

It started with the ringing of the alarm, clothes in a muddle, shoes that would not stay tied. Late again and forgotten a book. Trying hard to remember the number of things Mrs Smith had just asked us to do, leaned back to think, swung back too far and landed on the floor, everybody looking and laughing except Mrs Smith. Homework being written on the board, must try to write quickly before it is rubbed off. Bell goes, lesson ends, not too bad as lessons go, almost finished, hand aches, must have been pressing too hard on the pen again.

Lunch at last, must remember not to stand too near others in the queue, they think I bump into them on purpose; better wait till the last, hope there is still a choice left.

What lessons are there this afternoon? Check my timetable and bag, oh good PE my favourite lesson, I don't think, at least I remembered my kit. Final lesson geography, let's hope that I can finish my earthquake warning poster. Mr Peters says the idea is excellent, pity about the printing though, have to think of another way, still find typing on the keyboard difficult, could use one finger. Great, finished and looks good, no accidents or spillages.

DYSPRAXIA PROFILE	
Characteristics	Students may experience difficulties with the following: • Organising and co-ordinating fine and gross motor skills. • Controlling writing, drawing materials and tools. • Body awareness, moving about without bumping into things or invading other people's space. • Organising equipment and managing time. • Dealing with a sequence of events or instructions and following the timetable.
Potential impact of characteristics on learning	• Does not work well in an unstructured environment. • Finds organising equipment and managing time and finishing tasks hard. • Finds it difficult to sit still and is easily distracted. • Finds it difficult to transfer information from the board or to complete a complex sequence of instructions. • Becomes confused with locations and directions e.g. left and right, up and down, east and west etc. • Can become easily distressed and frustrated.
Possible supportive teaching and learning strategies	• Provide structure e.g. use pictures to represent what to do where and when. • Use of writing frames may be helpful (see Writing about development resource, chapter 4 resources on the CD). • Seating position in class should enable student to view teacher directly, close enough to hear and see instructions easily and avoid highly stimulating areas, e.g. next to window or door. • Consult physiotherapist and/or occupational therapist for appropriate sitting/writing positions or additional resource requirements. It often helps to have both feet resting on the floor, desk at elbow height and, ideally, with a sloping surface to work on. • Tasks and instructions need to be broken down into small steps. • Provide handouts, graphs, tape recorder etc. to support lesson content. • Help student to develop strategies to support organisational skills, e.g. checklists for resources required for specific tasks.

	• Organise buddies for oral work, who will also benefit from working with students and in turn be able to support them with oral work and be sensitive to their needs. • Check understanding verbally. • Use alternative methods to reduce handwriting requirements. • Supplement maps with pictures/photographs and oral interpretation.
Preparation and planning for outdoor learning	• Clear sequential instructions of events and task requirements provided prior to field study. • Structure for recording provided through writing frames. • Review of terrain in order to assess accessibility and/or additional support requirements.

www.dyspraxiafoundation.org.uk

Moderate Learning Difficulties (MLD)

The term 'moderate learning difficulties' relates to students who will be achieving considerably below their peers in most curriculum areas even when supported through long term intervention programmes and a flexible approach. Students with MLD may or may not have additional physical or medical conditions.

Main characteristics:

- difficulties with reading, writing and comprehension
- unable to understand and retain basic mathematical skills and concepts
- immature social and emotional skills
- limited vocabulary and communication skills
- short attention span
- under-developed co-ordination skills
- lack of logical reasoning
- inability to transfer and apply skills to different situations
- difficulty remembering what has been taught
- difficulty with organising themselves, following a timetable, remembering books and equipment

In geography these students can, if appropriate, be supported by modifying the geography programmes of study. QCA (2001[2]) state that staff can modify the geography programmes of study for students with learning difficulties by:

- Choosing material from earlier key stages
- Maintaining, reinforcing, consolidating and generalising previous learning, as well as introducing new knowledge, skills and understanding

- Focusing on one aspect, or a limited number of aspects, of the age-related programmes of study in depth or in outline
- Including experiences that let students at early stages of learning gain knowledge and understanding of geography in the context of everyday activities
- Helping students experience geography for themselves, at first by using a sensory approach to experience and investigate familiar places, and then by contact with different people

There is no requirement to teach geography at Key Stage 4. However, for many students, geography offers satisfying challenges both as a subject in its own right and as a means of developing skills in many other areas of the curriculum.

The full text relating to this document, 'Planning, teaching and assessing the curriculum for pupils with learning difficulties, Geography' can be found at www.nc.uk.net/ld/ge_content.html

Additionally, the Staffordshire Expanding Geography Scheme was developed by Diane Swift and Irene Corden to support curriculum planning for students with learning difficulties. The scheme is on the CD. It will be of value when planning curriculum experiences for students with moderate severe and profound learning difficulties.

Yasmin

Yasmin has moderate learning difficulties. She is in Year 7. Her attention-seeking and avoidance behaviour can prove disruptive in geography lessons as she often shouts out or continually talks to her peers. Her short-term memory skills are good and she can recall what she has been taught but she has difficulty in sequencing information. Lots of practice and repetition help her to consolidate new concepts in geography. This strategy is also helpful to her when learning how to use appropriate specialist vocabulary. She has a reading age of 9. In mathematics she is working confidently with numbers up to 50 but she has difficulty generalising from familiar to new or practical situations. She likes to write but needs considerable support. She enjoys using the computer and works confidently with the program Clicker 4, which supports her reading, writing and recording. Yasmin states that her favourite lessons are, 'when I can go outside and do things and work with my friends.' She enjoys geography field studies but requires lots of structure to organise materials, to help her to understand what is required of her and remain on task.

MODERATE LEARNING DIFFICULTIES PROFILE	
Characteristics	Students may experience difficulties with • Basic literacy and numeracy. • Expressive and receptive language. • Physical co-ordination. • Listening, attending and comprehension. • Memory. • Social and interactional skills.

Potential impact of characteristics on learning	• Poor reading, writing and numeracy skills. • Problems with processing, retaining and recalling information. • A limited vocabulary, immature use of language and grammatical structures. • Difficulties in understanding and responding to instructions and open-ended questions. • Poor listening and a short attention span. • Poor on-task and completion of task behaviour. • Poor organisational skills. • Immature social and emotional skills. • Disruptive behaviour through constantly distracting and talking to other students. • The use of a range of avoidance techniques. • Restless behaviour.
Possible supportive teaching and learning strategies	• Provide structure and routine to lesson format. • Break learning down into small steps. • Plan short tasks and varied activities, that include repetition of key concepts and ideas presented. • Differentiate in terms of content, resources and outcomes and use of a wide range of methods of communication including speech, images, pictures, charts, diagrams and symbols. • Support learning through the use of concrete examples, e.g. word lists, pictures, photos, symbols, sequencing prompts, shortening text, etc. • Repeat information and instructions in different ways. • Use of simple language and range of questioning techniques. • Rehearse with student what is to be recorded before they begin and use a range of structures to support reporting and recording, i.e. scaffolding, writing frames. • Provide environment and physical structure to reduce inappropriate and disruptive behaviours, e.g. sitting away from obvious distractions. • Identify clear rules and expectations. • Identify approaches to the management of disruptive behaviours, and avoid attention-seeking behaviours, and apply consistently. • Keep listening activities brief. • Catch them being 'good' and reward immediately.
Preparation and planning for outdoor learning	• Clear outline of activities and learning outcomes provided for student prior to field study. • Supported information provided through the visual resources that may also be used on the visit itself. • Clear identification of rules and behaviour expectations. • Structured recording format. • Group or paired activities.

The MLD Alliance c/o The Elfrida Society, 34 Islington Park Street, London N1 1PX
Website: www.mldalliance.com/executive.htm

Severe Learning Difficulties

Young people with severe learning difficulties may have a complex pattern of educational, social and emotional needs. Many have communication difficulties and/or sensory impairments in addition to more general cognitive difficulties. They can also have difficulties with mobility, co-ordination and perception. Some students may use signs and symbols to support their communication and understanding. Symbolic and spatial representation of course has great resonance with some of the key characteristics of geography. They will be achieving well below students of the same age and may remain within or below level 1 of the National Curriculum, or in the P scale range (see resources for chapter 6 on the CD) for most of their school life. You may find www.nc.uk.net/ld/ge_content.html helpful to inform your planning for these students. The access that such students have to mainstream settings varies between local education authorities. We hope that the following section will be supportive to colleagues currently supporting such students and will be useful preparation for those in settings who are yet to include students with severe learning difficulties. This section has been constructed drawing on experiences from teachers working in special school settings. There is informed dialogue and supportive liaison between special school and mainstream teaching teams. This underpins some highly effective practice.

Nicky

Nicky has severe learning difficulties. He is 11 years old and his education takes place both within a mainstream and special school. He spends half his time in each setting. His mainstream class has the full-time support of a teaching assistant.

Nicky uses a mixture of single words, phrases and simple sentences, accompanied by signing, to communicate. Often sensory experiences are provided for Nicky. These link to the learning objectives of the curriculum planned for the rest of the class. He enjoys joining in with small group activities. Differentiated resources, content, and learning outcomes enhance his access to the curriculum. He has a sight vocabulary of 50+ words and uses Widget symbols (www.widget.com) to aid both his reading and writing. Nicky has a basic number concept to 10 and can complete simple addition and subtraction, and applies these concepts in familiar practical and everyday situations.

Nicky can locate without support the main areas within the school and has been taught to use a simple pictorial and symbol map to find less familiar places that he does not access on a daily basis. He occasionally becomes confused with left and right although he uses his watch, always on his left wrist, to help him remember. He also has difficulties in identifying the position and location of objects in relation to himself, e.g. behind, by the side, near, far etc. He recognises basic differences and similarities in familiar but contrasting environments. In simple terms, and when supported through prior learning and careful questioning, he can identify an environmental issue and suggest ways to improve the situation, e.g. litter in school grounds.

SEVERE LEARNING DIFFICULTIES PROFILE	
Characteristics	Students may experience difficulties with • Personal independence and self help skills. • 'Learning to learn' and memory skills. • Expressive communication. • Developing basic skills. • Problem solving.
Potential impact of characteristics on learning	• Explicit learning programmes may be required in all areas of development. • Limited incidental learning, i.e. has difficulty picking things up as you go along and making connections. • 'Learning to learn' skills such as attention, concentration, on-task and group working skills may need to be specifically taught. • Limited vocabulary. • Limited use of language and comprehension. • Difficulties with problem-solving and responding appropriately to open-ended questions. • Difficulties in processing, retaining and recalling information and generalising knowledge, understanding and skills. • Poor reading, writing and mathematical skills.
Possible supportive teaching and learning strategies	• Individual educational programmes to address priority areas of learning. • The use of achievable tasks for each lesson or module of work. • Learning that is broken down into small steps. Allow time for students to process information and formulate responses. • Give a series of short, varied activities with each lesson and provide clear specific instructions about these. • Use simple language that is based on the student's understanding but continue to support and extend the student's communication, comprehension, vocabulary and use of language for a range of purposes. • Build on prior learning and introduce new concepts through concrete experiences whenever possible. • Set differentiated, practical tasks that link to the work of the rest of the class. • Objects, pictures and symbols relating to the place or theme being explored are used to support and supplement learning on a regular basis. • Devise doing activities that focus on specific areas of the geography curriculum, e.g. sensing different types of ground, soil, sand, grass, tarmac to appreciate places are both similar and different. • Identify key words and questions for geography, learn some relevant signs to communicate these appropriately to the student.

	• For each topic or theme, create a set of stimuli (and keep these noted down in an electronic format) in a box that appeal to different senses, e.g. things that smell, food from markets, things to touch, e.g. materials. • Use different recording methods, drawings, audio, or video recordings, use annotated photographs for each place and theme.
Preparation and planning for outdoor learning	• Provide a preview of location, activities, tasks to be completed and possible learning outcomes, e.g. use sequential geographical story of visit demonstrating activities, tasks and learning objectives, show video from previous visit made by other students or the teacher etc. • Working with student and parents, clearly identify organisation and resource requirements that need to be provided by the student, and support through use of the preferred communication system. • On the day provide resources to support the practical activities using student's preferred communication system. • Provide clear behaviour rules, boundaries and realistic expectations. • Identify possible additional adult, physical and medical support if required.

MENCAP, 117–123 Golden Lane, London, EC1Y 0RT Tel: 020 7454 0454
Website: http://www.mencap.org.uk

Profound and Multiple Learning Difficulties (PMLD)

Students with profound and multiple learning difficulties present a range of needs in addition to severe learning difficulties. They do not present a homogeneous group. In addition to very severe learning difficulties, students have other significant challenges to face, such as physical disabilities, sensory impairments or severe medical conditions. Students with PMLD require a high level of adult support, both for their learning needs and for their personal care.

They are able to access the curriculum through sensory experiences and stimulation. For some ideas as to how to use the sensory room in geography (see Appendix 4.2 and resources for chapter 4 on the CD). Some students communicate by gesture, eye-pointing or symbols, others by very simple language. They may work within P1–P4 for most of their school career (see chapter 6 resource relating to P scales on the CD). The P scales provide small, achievable steps to monitor progress. Some students will make no progress or even may regress because of associated medical conditions.

To support you with your curriculum planning for this group you may wish to refer to the Staffordshire Expanding Geography Scheme (on the accompanying CD).

As with the previous section relating to students with severe learning difficulties, the access that students with profound and multiple learning difficulties have to mainstream settings varies between local education authorities. Again we hope that the following section will be supportive to colleagues currently supporting such students and will be useful preparation for those in settings who are yet to include PMLD students. This section has been constructed drawing on experiences from teachers working in special school settings. Informed dialogue and supportive liaison between special school and mainstream teaching teams underpins some highly effective practice.

Alice

Alice's Statement of Special Educational Needs identifies that she has 'difficulties in general learning skills, visual, communication, interaction, mobility and independence skills'. It is also noted that Alice has a visual impairment and intermittent conductive hearing difficulties. Assessment of Alice's level of achievement shows that she is presently working at a P2 (ii)/P3 (i) level in the core subjects.

Alice can present challenging behaviour, such as hair pulling, grabbing, and pinching people who are in close proximity. This is often the result of unexpected changes or confusion. This behaviour has proved a useful strategy by which Alice can control invasions into her personal space. Alice has recently become an independent walker and her confidence in her ability to move about within familiar areas is increasing.

Alice attends a special unit within a mainstream setting. She is a student within a class of 9, where the class teacher is supported by 2 teaching assistants. Alice has 5 hours of additional individual support from a key worker each week. The advisory teacher for the sensory impaired visits each half term, to support the work with Alice.

Alice's developing understanding of her school day is helped by a holistic approach to communication, incorporating the use of song, sign, objects of reference, and familiar daily classroom routines.

Both the human and physical environments are structured to support students with sensory impairments.

All staff use a personal signal to alert Alice to who they are. They approach Alice in an unhurried and standardised way, in order to lessen her fear of the unexpected and to develop positive anticipation of human interaction. All staff members have an awareness of Alice's mobility needs. Signals are used to support Alice's understanding of changes activity or location.

Alice's sensory impairment means that she may receive an incomplete or skewed view of the world and many aspects of that environment may be unpredictable. In order to provide security for Alice and to support her understanding of place and function, the environment is structured, e.g. a personal work table in a specific and unchanging area of the classroom, and a relaxing area with access to favourite music etc. Additional communication approaches, physical, sensory and verbal cues and a consistent approach by all staff provide Alice with the growing confidence to explore her immediate and wider environment.

PROFOUND AND MULTIPLE LEARNING DIFFICULTIES PROFILE	
Characteristics	Students may experience difficulties with • Movement, physical disabilities may require specialised physical resources, these may include a wheelchair, standing frame, side liner. • Sensory impairment. • Complex medical conditions. • Inappropriate behaviour. • Communication. • Retaining experiences. • Learning beyond that associated with the early stages of development in all areas.
Potential impact of these characteristics on learning	• Require a high level of adult support to access learning, their environment and meet personal, physical and medical needs. • Limited movement that may seriously restrict the student's ability to physical access, react to, and impact on their environment. • Visual, hearing or dual sensory impairment. • Physical and medical conditions may impair sensory functions. • Difficulty in processing information. • Relies on observant and vigilant staff to respond to personal methods of communication, i.e. movement of left hand, vocalisation.
Possible supportive teaching and learning strategies	• Individual educational programmes designed by a multi-professional team that provide an holistic approach to teaching and learning. • Learning that is broken down into small, achievable steps. This may well focus on the development of the student's sense of personal geography. • Multi-sensory dimension to teaching and learning, e.g. providing sight, sound, taste experiences relating to places, themes or concepts that the class is addressing. • Focus on real experiences in a variety of settings to support a sense of place. • Explicit teaching of 'learning to learn' and social skills. • Revisit learning and extend in different contexts and settings. • Revisit locations and experiences in school through multi-sensory activities. • Consistent approach by all staff. • Supportive and flexible learning environment and timetable to provide access to inclusive as well as specialist learning opportunities or settings.
Preparation and planning for outdoor learning	• Be creative in reviewing what field studies may provide the required learning experiences and opportunities for the students. These may not involve great distances or difficult terrain.

	• Pre-visit by the teacher to assess physical access, facilities, staff requirements, medical support and any additional risks that the visit may pose. • Communicate with the student throughout the visit about what is happening immediately and what will happen next. • Use aspects of the environment to extend students' knowledge of the location, e.g. sounds, smells etc. of a busy market in contrast to the smell, sounds of a quiet wooded area; use of the school grounds to explore different weather conditions, cobbled path, smooth path, rough terrain.

MENCAP, 117–123 Golden Lane, London, EC1Y ORT Tel. 020 7454 0454
Website: http://www.mencap.org.uk

Fragile X

Fragile X is a complex genetic condition and is one of the most common known causes of inherited learning difficulties. It occurs because of an abnormality on the long arm of the X chromosome that can cause a variety of intellectual, behavioural and physical differences. Fragile X affects both males and females, although it is often the case that males exhibit more severe symptoms than females. Cognitive and learning abilities may range across the whole spectrum.

Main characteristics:

- delayed and disordered speech and language development
- difficulties with the social use of language
- articulation and/or fluency difficulties
- verbal skills better developed than reasoning skills
- repetitive or obsessive behaviour such as hand-flapping, chewing, etc.
- clumsiness and fine motor co-ordination problems
- attention deficit and hyperactivity
- easily anxious or overwhelmed in busy environments

Saul's Story

Saul is 15 years old and attends an integration unit on a mainstream school site. He has a great sense of humour, enjoys taking responsibilities and is a great favourite with adults. Saul appears to have no friends of his own age and has little contact or interaction with his peers. He is active and impulsive and often rushes into situations without thinking or being aware of what the consequences may be. Saul has a very good vocabulary and usually speaks in short phrases and sentences. He has a number of very amusing learned phrases that he uses in the correct context. He finds it difficult to answer direct questions and often gives answers that appear unusual or form part of his learned repertoire. Although Saul demonstrates that he enjoys verbal praise he appears to find it hard to accept and often responds by giggling or becoming extremely embarrassed. Saul becomes extremely anxious within social and group situations and will begin to chatter to himself using well-known phrases or instructions that he has been given and repeat specific hand, head and body movements. These activities appear to bring the levels of anxiety down and have a calming effect. Lots of other situations throughout the school day raise his anxiety levels including demanding class work, busy and noisy environments and changes in routines. He does not display any challenging behaviour within the school setting. Saul finds a structured environment, a visual and word timetable, clear instructions and adult modelling supportive to his learning.

In English and Mathematics he is working at Level 1a and in other subjects P8 of the National Curriculum.

FRAGILE X PROFILE	
Potential impact of characteristics on learning	• Difficulties in attending, remaining on task and transferring from one task to another. • Difficulties with problem-solving, thinking and reasoning skills and abstract learning tasks. • Sees tasks as whole rather than the individual. • Is likely to be a visual learner. • Social anxiety often inhibits a student's ability to communicate affectively. • May find cause and effect and open-ended questions difficult. • May present difficulties in social skills and adaptation of such skills in new settings. • May present repetitive and sometimes unusual behaviour, e.g. hand flapping, and demonstrate over-activity, social anxiety, inflexibility and an insistence on routine. • Susceptibility to stimuli overload and busy environments.
Possible supportive teaching and learning strategies	• Group and social skills need to be taught explicitly. • Provide timetables and lists to support sequences, transfer of learning and change. • For many students with SEN we tend to break tasks down into smaller steps. However, for some students with Fragile X it may be more appropriate to use the opposite approach, i.e. demonstrate the whole and then the individual components. • Provide concrete examples and/or model task or activity whenever possible.

	• Use varied activities and tasks and keep them short. • Give clear indications throughout the structure of the task about what is expected, what has to be done and when it is completed. • Use of alternative ways of recording, e.g. computer, adapted worksheets that are free from clutter. • Provide a seating arrangement to avoid busy areas of the classroom and which allows the student not to feel crowded or that their personal space is being invaded, e.g. to the front and side. • Agree with the student strategies to manage over-stimulation, frustration etc. • Provide visual information (including maps and photographs) that are simplified to eliminate clutter or excessive stimulation. • Provide objects, artefacts and visual resources to support learning. • Whenever possible relate learning to the student's everyday experiences. • Provide visual resources to help the sequencing and organising of tasks, e.g. arrows, numbers etc.
Preparation and planning for outdoor learning	• Provide structured materials, worksheets and visual resources to clearly identify the purpose of each task, the tasks themselves and the sequence of activities and expected outcomes. • Make learning objectives clear. • Use supportive or co-operative groupings. • Make expectations and rules for behaviour clear.

Fragile X Society, Rood End House, 6 Stortford Road, Dunmow, CM6 1DA
Tel: 01424 813147 (Helpline) Tel: 01371 875100 (Office) Email: info@fragilex.org.uk
Website: http://www.fragilex.org.uk

Down's Syndrome (DS)

Down's syndrome is a genetic condition that can be caused by the presence of an extra chromosome 21. Students with Down's syndrome, while sharing certain physical characteristics and a specific learning profile, will also inherit their own family's looks and characteristics. They will have their own talents and aptitudes. The majority of students will also have some degree of learning difficulty that may range from mild to severe. They may have additional sight, hearing, respiratory and heart problems.

Main characteristics:

- delayed motor skills
- take longer to learn and consolidate new skills

- limited concentration
- difficulties with generalisation, thinking and reasoning
- sequencing difficulties
- stronger visual than aural skills
- better social than academic skills

Susan's Story

My name is Susan, I am nearly 16 and I have Down's syndrome. I don't suffer from Down's syndrome as many people seem to think, I was born that way and I feel great about myself. I know that I have an extra chromosome and that I have some learning difficulties but so do lots of my friends and they don't have Down's syndrome. I also have friends with Down's syndrome and they are all different too.

I attend my local secondary school. I am good at reading but not so good at maths and I use a computer to help me with my writing. Sometimes people who don't know me well find it difficult to understand what I am saying but my family, friends and teachers don't have any problems at all.

DOWN'S SYNDROME PROFILE	
Potential impact of characteristics on learning	• Poor co-ordination that may impact on physical activities and writing skills. • Smaller vocabulary, articulation problems, difficulties in learning the rules of grammar. • Processing, understanding, assimilating and responding to spoken language. • Difficulties in understanding complex verbal instructions. • Poor problem-solving and generalisation of knowledge, understanding and skills. • Difficulties in processing, retaining, organising, consolidating and sequencing information. • Difficulties in recording information, e.g. selecting, organising and sequencing relevant information. • Short concentration span. • Sensory functions.
Possible supportive teaching and learning strategies	• Investigate additional resources to support physical aspect of writing and alternative approaches to recording. • If appropriate seek advice of other professionals, i.e. physiotherapist, occupational and/or speech therapist. • Check understanding through questions. • Build on prior knowledge, reinforce abstract concepts with supplementary visual resources and artefacts relating to the place, theme or concept. • Reinforce new skills and generalisation by using a range of methods in different settings and by offering opportunities for repetition.

	• Support note taking by highlighting key information, or by using scaffolds and writing frames. • You may also find the use of key word banks, visual resources and sentence sequencing cards helpful to support recording. • Provide a supportive structure by teaching routines and rules explicitly and warn in advance any changes that may occur. • Ensure students' position in class and resources take into consideration visual or auditory difficulties. • Reinforce speech through visual resources and/or signs. • If required set differentiated tasks linked to the work of the class.
Preparation and planning for outdoor learning	• Clear sequential instructions of events and task requirements provided prior to field study. • Structure for recording provided through writing frames or scaffolds. • Promote paired or group work. • Review of terrain in order to assess accessibility and/or additional support requirements.

The Down's Syndrome Association, Langdon Down Centre, 2a Langdon Park, Teddington, TW11 9PS Tel: 0845 230 0372 Email: info@downs-syndrome.org.uk Website: http://www.downs-syndrome.org.uk

Behavioural, Emotional and Social Difficulties (BESD)

Students with behavioural, emotional and social difficulties cover the full ability range found within the mainstream setting. A student will be diagnosed as having special educational needs when the severity of the behavioural, emotional or social difficulties forms a barrier to learning that is significantly greater than most of their peers. In addition, as part of the diagnosis the character, frequency, persistence, severity and increasing effect of the behaviour on the learning of themselves and others will be taken into consideration. Individual behavioural programmes for students with BESD will be developed on the basis of each student's specific need and with advice from an educational psychologist and, where appropriate, professionals from other agencies. Therefore the teaching strategies suggested below only identify general classroom management strategies.

Main characteristics:

- inattentive, poor concentration and lack of interest in school/school work
- easily frustrated, anxious about changes
- unable to work in groups
- unable to work independently, constantly seeking help
- confrontational – verbally aggressive towards students and/or adults
- physically aggressive towards students and/or adults
- destroys property – their own/others
- appears withdrawn, distressed, unhappy, sulky, may self-harm
- lacks confidence, acts extremely frightened, lacks self-esteem
- finds it difficult to communicate and to accept praise

Sally

Sally is 13 years old and attends the local high school. She has experienced a very disruptive family life and has been in and out of care since the age of 4. At this time she lives with foster parents. Her attendance at school is poor. When she does attend she is often late and on entering a class will knock objects off the desks of other students and shout across the room to her friends. She refuses to listen in class or complete tasks set, and when requested to work she will either ignore the teacher or respond with abusive or threatening language. This behaviour is exhibited with all teachers at some time.

Individual behaviour management plans (IBMP) have been revised and adapted where appropriate in co-operation with the educational psychologists and foster parents. Teachers have worked closely together to share information and provide mutual support. The frequency of the outbursts has decreased over the past 3 months.

BEHAVIOURAL, EMOTIONAL, SOCIAL DIFFICULTY PROFILE	
Potential impact of characteristics on learning	• Demonstrates a lack of interest in school or learning. • Underachieves. • Poor attendance. • Finds it difficult to concentrate and complete tasks, works slowly. • Often displays inappropriate behaviour that disrupts the lesson, is harmful to self and others or is extremely isolated, withdrawn, distressed, unresponsive, or shows unusual behaviour. • Demonstrates mood swings and is easily frustrated or emotional. • Finds it difficult to work co-operatively with others, share equipment or space. • May find it difficult to accept rules or what is perceived as authoritarian figures and may resort to confrontation or argument. • Failure to follow instructions. • Poor organisation. • May destroy own work or that of others. • Finds it difficult to communicate, trust, interact and form relationships with adults and peers.
Possible supportive teaching and learning strategies	• Differentiate work, content and outcome to meet student's achievement levels. • List main learning objectives, ideas, concepts and tasks to be completed at the beginning of lessons. Make clear some of the big ideas in geography to the students. • Provide a calm structured environment with clear routines and procedures. Use the stimulation of geography, about real people and real places. • With the school's behaviour policies and procedures clearly identify behaviour expectations, rules, routines and responsibilities. • Have a small number of important rules and apply consistently rather than many that are unmanageable. • Identify clearly the consequences of inappropriate behaviour and implement fairly. • Implement, monitor and review IBMP continually. • Recognise and acknowledge positive behaviour in a way that is appropriate to the student and identify ways to enhance their self-esteem. • Tell students what you want them to do, e.g. 'I need you to . . .', rather than ask, e.g. 'Will you . . .', which provides the opportunity for the negative response 'No!' • Use strong, positive 'I' statements, e.g. I care, I'd prefer, I'd be happier if . . . • Choose appropriate times for discussion and negotiation, e.g. not in full view of others. • If confrontation occurs, stay calm, lower your voice.

	• Do not enter into in-depth confrontations or discussions, deflect student, suggest next step, give choice. • Focus on behaviour and not the student.
Preparation and planning for field studies	• Make behavioural expectations and rules explicit. • Identify clearly the dangers in specific locations. • Ensure adequate staffing.

SEBDA is the new name for the Association of Workers for Children with Emotional and Behavioural Difficulties. Website: http://www.awcebd.co.uk

Attention Deficit Disorder/ Attention Deficit Hyperactivity Disorder (ADD/ADHD)

Attention Deficit Disorder is the term used to describe a neurological based dysfunction which affects performance, self-regulation and impulsivity. Some students may also present hyperactivity. Students demonstrate a normal range of intellectual ability but the ADHD may in itself cause barriers to learning. All students are different and will present individual patterns of behaviour. Many will display some if not all of the characteristics listed below to a degree.

Main characteristics:

- difficulty in following instructions and completing tasks
- easily distracted by noise, movement of others, objects attracting attention
- often doesn't listen when spoken to
- fidgets and becomes restless, can't sit still
- interferes with other students' work
- can't stop talking, interrupts others, calls out
- runs about when inappropriate
- has difficulty in waiting or taking turns
- acts impulsively without thinking about the consequences

Callum's Story

My name is Callum and I am 13 years old and have a diagnosis of ADHD. I take medication to help me to control my feelings and to keep me calm. I try to make friends but I find it difficult to do what they say, keep to the rules or wait my turn; then they don't want to know me and tell me to go away. Then I hit or kick; my feelings frighten me sometimes.

Sometimes I get very angry with my parents and teachers because they are always telling me what to do or think I am not listening because I am 'jigging' about, but for me this is how I listen the best. They don't understand how difficult it is to sit still and then listen to what they say. I suppose they are alright most of the time, it's just that I get so excited about things and I can't wait to try them or share what I know or have found out. My Mum says, 'I do or say things before I put my mind in gear and that is why I get into lots of trouble.' Another thing that drives my family and teachers mad is that I am always losing or forgetting things, what they don't realise is that this makes me as angry as they are.

I know that I am good at lots of things at school and have a very good memory but it is not always easy for me to listen carefully for a long time. Sometimes my thoughts jump from one thing to another and suddenly everything has moved on in a lesson and I am lost, have missed what I have been asked to do or answer the wrong questions. Some teachers think I do this on purpose, I don't honestly.

ADD/ADHD PROFILE	
Potential impact of characteristics on learning	• Talkative, energetic, enthusiastic, inquisitive, impulsive and adventurous. • Easily distracted, cannot maintain attention, refocus on tasks, forgetful, disorganised and has poor attention to detail. • Slow to complete tasks/work set. • Acts and speaks without thinking, interrupts, may have low frustration and tolerance threshold. • May appear dreamy and distant or restless, fidgety, talkative, difficulty in remaining in seat, waiting turn. • Will continue to discuss or argue and will not let issues drop. • Misunderstands social situations, demands own way. • Poor co-ordination, clumsy, written work may be messy. • No concept of structuring environment, belongings, school work or projects, loses things, unaware of time. • May demonstrate mood swings. • May experience specific learning difficulties or underachieve.
Possible supportive teaching and learning strategies	• Accept the student for who they are and acknowledge their good points. • Develop a trusting and positive relationship with the student that promotes the student's self-esteem. • Link the geography curriculum to real life experiences and real life applications. • Structure environment and reduce clutter and sit student away from distractive or busy classroom areas. • Make sure that maps and other graphical representations are clear and that only essential information is included. • List main learning objectives, ideas, concepts and tasks to be completed at the beginning of lessons. Make clear how place studies relate to some of the big ideas. • Break tasks down into small steps.

	• Give clear directions orally and visually and where possible provide a model of what to do. • Provide breaks within longer tasks and/or alternative lesson activities. • Find alternative methods other than tests to identify what students know or have learnt, such as self-assessment and peer review. • Provide opportunities for the student to move about legitimately, i.e. take a message, vary activities within the lesson. • Help students to use written lists that identify personal responsibilities, resource requirements, tasks, activities etc. • Have clear and realistic behaviour expectations that are expressed in positive terms. • Tell the student well in advance if changes are to take place. • Do not enter into in-depth confrontations or discussions, rather deflect student to previous activity that they have found calming. • Provide visual resources alongside verbal explanations and lists to support the organisation of tasks. • Repeat and paraphrase key concepts in a variety of ways. • When asking questions allow student thinking time before requesting a response.
Preparation and planning for outdoor learning	• Make learning objectives clear. • Use co-operative groupings. • Make behavioural expectations and rules behaviour explicit.

ADD Information Services, PO Box 340, Edgware, Middlesex HA8 9HL
Tel: 020 8906 9068 Website: www.btinternet.com/~black.ice/addnet/

COMMUNICATION AND INTERACTION

Speech, Language and Communication Difficulties (SLCD)

Speech and language are the means through which we communicate or share our thoughts, ideas and emotions. Language is a shared set of rules or system of rules through which we communicate using speech, reading or writing. The development of language is a complex process that relies not only on the development of organs or muscles that enable or support the production of speech but also the ability to use (expressive) and understand (receptive) language. Understanding of the nuances in verbal and non-verbal communication and social interaction enables us to add to or change the meaning of the words that we use. Difficulties in the development of speech, language and communication may involve any aspect of this process, cover the full continuum of severity and may last for a short or long period of time. If the difficulties are not related to any other SEN they are identified as specific language difficulties. It is not difficult to anticipate the impact of specific language difficulties on the learning, social and emotional development of any student.

Main characteristics:

General

- specific disability in some aspect of speech, language or communication
- poor interactional and social skills
- emotional difficulties and poor self-esteem
- poor attention and listening skills
- assimilating and generalising new concepts and skills and the associated language

Expressive language

- phonology – delay in, or disorder of, speech sound systems
- semantics – difficulty in conveying meaning to others
- pragmatic – difficulties in understanding the rules of communicative interactions
- syntax and grammar – limited use of complex sentence structures

Receptive/comprehensive difficulties

- semantics – difficulty in understanding the meaning of language
- grammar and syntax – difficulties in understanding specific sentence structure

- sequencing – difficulties in the sequencing and organisation of words in sentences and sequencing information
- poor language memory and retention skills

Chantelle's Experience of Transition to High School

My name is Chantelle and I recently started at secondary school. The teachers from both my primary and secondary school provide lots of help. I went with my class and friends to visit the school in the summer term. The teachers from both schools met with my parents and me.

It is very different at my new school. It is very big and, at first, even with my map and help from my friends I still could not find my way around all the rooms. We have to move from one class to another for most of the lessons. The things that really worry me are all the new rules, all the things I have to bring with me each day, the amount of work we have and how quickly we move on to new things in each subject. I also don't feel that other people in the class or every teacher understand my problems.

I suppose I should tell you a little about my difficulties, but that is really hard. Perhaps the first thing to say is that I often get words in a muddle, this is when I am talking and writing. Sometimes I know that it is the wrong word but can't remember the right one. Sometimes I don't know that I have said the wrong word until my friends laugh. I also find it difficult to put what I want to say in the right order when I am talking and writing. Now that I am at the high school there seem to be a lot of new words to learn in each subject. I also have to learn to spell them too which is very difficult when you can't remember the word in the first place. Reading and writing and of course talking are not my best skills, but there they are in every lesson so I will just have to get on with it. I do find it hard to listen and understand and take notes but the teachers are very good and I use a tape recorder which helps me to go over them later with my Mum. Listening and talking when we work in small groups is also hard but we have a special way of working and so I get better every day. One of the best things is that I now have my own lap top which I take everywhere with me. It has special programmes to help me and I am also much quicker with my work.

COMMUNICATION AND INTERACTION PROFILE	
Potential impact of these characteristics on learning	General • Poor literacy skills. • Difficulty in developing and maintaining friendships and understanding of social rules. • May become anxious, frustrated or may under achieve. • Poor self-confidence and self-esteem. • Intense concentration leads to student being tired and exhausted by the end of the day. • Difficulty in organising self, environment and materials and following timetables, class routines or rules. Phonological • Poor articulation and intelligibility of speech. • Limited use of more complex sentence structures. • Difficulties in the use of phonics impacting on reading and literacy skills.

	Semantics • Difficulty in communication or conveying meaning to others. • May use incorrect vocabulary. • Information lacks detail. • Loses focus or subject of conversation. Pragmatics • In communicative exchanges may not maintain eye contact, take turns, use language appropriately for purpose or settings. • Speech errors, e.g. irregular plurals, tenses. • Unable to sequence, plan or organise information or ideas in a coordinated and logical way. • Difficulties in remembering and retrieving vocabulary, or may use incorrect label, e.g. use cup when they mean spoon.
Possible supportive teaching and learning strategies	• Work with students, parents and speech therapists to develop individual education programme based on student's individual need or identify specific teaching strategies and resources. • Use visual resources, pictures, symbols, computer programmes to support communication and language skills. • Support lessons and introduction of new subject specific concepts and vocabulary through concrete experiences, and the use of visual resources. • Explicitly teach social, interactional, communication and co-operative working skills as normal part of lesson structure and group work. • Use lists, subject vocabulary banks, sequencing cards, adapted work sheets, augmentative writing systems etc, to support organisational, speech, language and recording skills and activities. • Provide subject specific vocabulary work banks, visual and topic uncluttered displays/pictures with vocabulary and label or colour code equipment and resources. • Model task and activities and use of equipment. • Use tape recorded descriptions of frequently used maps and photographs. • Use range of strategies to problem solve, explore ideas and plan, review, record what they know and have learned, e.g. mind maps, computer overlays etc. • Alert the student that you are communicating with them, directing information by calling out their name. • Clearly outline lesson objective, activities and learning outcome and you may find that repetition at each stage of the process is helpful.
Preparation and planning for outdoor learning	• Clearly outline objectives, activities and learning outcomes prior to visit and discuss with student possible additional resource and recording requirements.

	• Support organisational skills through schedules and timetable. • Plan for group activities and review communication and language skills to be used. • Make explicit the rules and expectations.

ICAN, 4 Dyer's Buildings, Holborn, London EC1N 2QP Tel. 0845 225 4071
Email info@ican.org.uk Website: http://www.ican.org.uk
AFASIC, 2nd Floor, 50–52 Great Sutton Street, London, EC1V ODJ
Tel. 0845 355 5577 (Helpline) Tel. 020 7490 9410 Fax. 020 7251 2834
Email: info@afasic.org.uk Website: http://www.afasic.org.uk

Autistic Spectrum Disorder (ASD)

Students with ASD have difficulties in social interaction, communication and areas that involve imagination and flexibility of thought. This is known as the 'Triad of Impairments'. They may be quite creative in ways that revolve around special but restrictive interests. They may display quite elaborative routines or the need to do things in a particular way. They may also have difficulties with sensory overload or have repetitive sensory behaviours that are pleasurable but may present barriers to learning. They may be very disorganised and clumsy or very obsessive about things being in the order or position that they need them to be in. They may be quite assertive in these demands or may well become socially isolated and removed. Students with ASD cover the full range of ability, and the severity of the impairment varies widely. Some students also have learning disabilities or other difficulties. Four times as many boys as girls are diagnosed with ASD.

Main characteristics:

- **Social Interaction**

Students with ASD find it difficult to understand social behaviour and this affects their ability to interact with children and adults. They do not always understand social contexts. They may experience high levels of stress and anxiety in settings that do not meet their needs or when routines are changed. This is particularly significant when planning out of classroom activities, since this change can lead to inappropriate behaviour. The use of social stories that rehearse situations prior to the event are helpful here. A department may wish to create a bank of such stories that consider appropriate behaviour in fieldwork situations.

- **Social Communication**

Students with ASD have difficulty understanding the communication of others and in developing effective communication themselves. Often their understanding of the use of non-verbal and verbal communication is impaired. They have a literal understanding of language. Many are delayed in learning to speak, and some never develop speech at all.

- **Social Imagination and Flexibility of Thought**

Students with ASD have difficulty in thinking and behaving flexibly which may result in restricted, obessional or repetitive activities. They are often more interested in objects than people, and may have intense interests in one particular area, such as trains or vacuum cleaners. Students work best when they have a routine. Unexpected changes will cause distress. Some students with Autistic Spectrum Disorders have a different perception of sounds, sights, smell, touch and taste and this can affect their response to these sensations.

John's Story

I have ASD which means I have special skills in a few areas. I know I have to do things in certain ways and then I feel calm. I don't like school that much but I do like geography. This is because I can look at maps and think about the motorways. Sometimes kids in my class shout at me because I like to make my special noise even though I do not always know when I am making it. I need things written down for me to understand what I have to do. It takes me longer to work out what people say to me and sometimes teachers get angry because they think I am not listening or refusing to do what I am told. I love clocks and how they work, sometimes I can talk about clocks in class because we talk about time zones. I like to go to my nan's. We go along three different motorways and I calculate how many lamp-posts there are on each and work out the mileage average between them. When I get back, I look at the maps on the walls at school and it makes me happy. I don't like people especially girls asking me about their clothes because I don't want to look at them and I know I can't get the answer right and it makes me sweat. I like to work on my own. I like working on the computer because it does not get angry.

AUTISTIC SPECTRUM DISORDER PROFILE	
Potential impact of characteristics on learning	• Can sometimes be so distant in lessons that the student does not hear instructions. • Does not always understand conceptual expressions or phrases, and misunderstands what is being asked of him/her. • The student needs things to be concrete or visual to make sense so doesn't always understand questions. • Needs teachers to call their name for them to realise they are included in an instruction. • Often shows signs of stress when concentrating and is agitated.
Possible supportive teaching and learning strategies	• Give advance warning of any changes to usual routine, e.g. change of classroom, change of teacher etc. • Needs worksheets to be very obvious in what they demand. Link the task with the learning objectives. • Liaise with parents, they may have some useful strategies. • Avoid asking direct questions but ask the student to write answers down.

	• Give individual instructions, using the student's name. • Avoid using too much eye contact as this can cause distress. • Use real examples of places within their experience. • Use as many real life experiences as possible. • Provide visual resources in class: objects, artefacts, pictures, picture dictionaries. • If possible put student in a small group if group work is necessary. • Use computers. Students with autism can have unusual fixations on parts of objects. They may focus on objects as though through a tunnel. When using a computer this can mean that they are able to focus totally on the screen and to block out all distractions around them. Using a computer can provide a secure comfortable environment as they feel more in control of their surroundings. They are less likely to fail and they can choose whether or not to communicate with others.
Preparation and planning for outdoor learning	• Give clear instruction on tasks to be completed. • Use a digital camera or video to record appropriate behaviour. Put the photos into a book and add a sentence to each one: 'Here I am using a clipboard carefully'. Read through this book before each field trip or out of classroom experience. This is an example of a social story activity. • Give clear guidance on suitable behaviour whilst out of school, extend through the use of social stories. • Put in group where social skills are low key.

The CD provides additional information on Autism, including an additional case study.

The National Autistic Society, 393 City Road, London EC1V 1NG Tel: 0845 070 4004
Helpline (10am–4pm Mon–Fri) Tel: 020 7833 2299 Fax: 020 7833 9666
Email: nas@nas.org.uk Website: http://www.nas.org.uk

Asperger's Syndrome

Asperger's Syndrome is a disorder at the able end of the autism spectrum. People with Asperger's Syndrome have average to high intelligence but share the same Triad of Impairments. They often want to make friends but do not understand the complex rules of social interaction. They have impaired fine and gross motor skills, with writing being a particular problem. Boys are more likely to be affected – with the ratio being 10:1 boys to girls. Because they appear 'odd' and naïve, these students are particularly vulnerable to bullying.

Main characteristics:

- **Social Interaction**

Students with Asperger's Syndrome want friends but have not developed the strategies necessary for making and sustaining friendships. They find it very difficult to learn social norms. Social situations, such as lessons, can cause anxiety.

- **Social Communication**

Students have appropriate spoken language but tend to sound formal and pedantic, using little expression and with an unusual tone of voice. They have difficulty using and understanding non-verbal language, such as facial expression, gesture, body language and eye contact. They have a literal understanding of language and do not grasp implied meanings.

- **Social Imagination**

Students with Asperger's Syndrome need structured environments, and to have routines they understand and can anticipate. They excel at learning facts and figures, but have difficulty understanding abstract concepts and in generalising information and skills. They often have all-consuming special interests.

Mary

My name is Mary, I am 15 years old and have a diagnosis of Asperger's Syndrome. When I was small I could never understand other children's behaviour. I would watch them playing the same favourite game many times but could never understand the rules and therefore could never join in. As I grew up I became even more confused about social interactions, although I have tried to learn the rules they are always changing and often appear illogical. Unfortunately I am unable to interpret non-verbal communication and only hear words, therefore I have great difficulty following a conversation, listening, being tactful, taking hints, making small talk and knowing what is acceptable to say. Consequently, I am always upsetting people, and have few friends. Other people also think I am rude or don't listen but once I am focused on something I cannot just think about something else.

ASPERGER'S SYNDROME PROFILE	

Potential impact of characteristics on learning	• May interpret language literally.
	• May speak fluently but have difficulties with participating in effective two-way communications, e.g.
	  • taking turns in conversation;
	  • does not understand and is therefore unable to adapt to responses of others who are listening to them;
	  • may continually talk about own special interests and/or may make comments that could be seen as offensive;
	  • appears insensitive to the feelings of others.
	• Language may be over precise or too literal and misunderstanding may occur over multiple levels of meaning, humour and metaphors.
	• May have a desire to be sociable but still have difficulties in understanding non-verbal signals or cues including facial expressions.
	• Difficulties in forming and maintaining social relationships and interactions.
	• May develop 'special interests' that may become obsessional.
	• Regulating of personal emotions may be difficult.
	• Often dislikes change, and therefore develops daily routines.
	• Good memory and retention of facts but may have difficulties with abstract thoughts.
	• Attention, concentration and on-task behaviour may vary considerably.
	• Recognition of the written word may be good but there may be difficulties with comprehension especially at a high level of literacy where interpretation of subtle meanings is required.
	• Written work may be repetitive and not focused.
	• Academic work may vary in quality depending on how meaningful or interesting the subject appears to the student.
	• May not always complete work.

Possible supportive teaching and learning strategies	• Understanding that the student's perspective of the world will differ from that of other students. • If possible raise awareness of peers to the needs of the student to avoid possible misunderstandings related to social situations and communication. • Interactive and group skills may need to be explicitly taught. • Use group strategies that highlight the student's strengths thus supporting peer acceptance. • Identify clear rules and expectations in relation to work and behaviour. • Provide a consistent approach supported by routines but introduce and teach an approach signal to prepare the student for change.

	• Agree strategies with the student if stress or frustrations overcome him/her. • While recognising the need for students to discuss their 'special interests', manage the possibility of it becoming all pervading by allocating a specific time when this type of discussion may take place in geography. • When possible, link their interest to subject being taught, e.g. when exploring landforms, a student obsessed by mountain ranges might be asked as part of a group task to research this particular landform and its use. • Add additional explanations or clarity when introducing new concepts or ideas, especially if they are abstract, and check for understanding. • Promote completion and speed of written work through the use of structured worksheets and the use of computer programs. • Use supplementary visual resources, pictures maps and photographs relating to the place, concept or theme being discussed. • Frequently direct questions towards the student to engage them in lesson but allow time for response. Alert them to this by using their name.
Preparation and planning for outdoor learning	• Provide structure to all aspects of the field trip, i.e. pre-planning, preparation, organisation, task requirements and completion, identify or recording outcomes and learning. • Give clear instruction on tasks to be completed. • Give clear guidance on suitable behaviour whilst out of school, extend through the use of social stories.

The National Autistic Society, 393 City Road, London EC1V 1NG Tel: 0845 070 4004
Helpline (10am–4pm, Mon–Fri) Tel: 020 7833 2299 Fax: 020 7833 9666
Email: nas@nas.org.uk Website: http://www.nas.org.uk

SENSORY AND/OR PHYSICAL IMPAIRMENT

Visual Impairment (VI)

There is a wide range of visual impairment in both type and degree. These include conditions of short- and long- sightedness that can be successfully corrected through glasses to students who have virtually no useful sight and are registered as blind. The majority of students who are registered as blind have some sight but the degree will vary considerably depending on the cause, e.g. a person with tunnel vision may be able to read but have difficulties with mobility while for others the situation will be reversed. Students with visual impairment may cover the whole ability range.

Main characteristics:

These will vary from student to student depending on the affect of their specific visual impairment but may involve the following:

- difficulties with mobility
- limited spatial experience leading to poor spatial awareness
- difficulties with social and communication skills
- become tired and fatigued after periods of intense concentration.

Louise

Louise is a Year 7 student with a visual impairment. She is currently placed at School Action Plus on the SEN Register. She has damaged optic nerves and reduced vision. Her sight is particularly poor at a distance and bright light causes her discomfort and further reduces her vision. Louise is a very bright student who always seemed to deal with her impairment in a matter of fact way, though at high school she has become more conscious of her impairment and this has affected her self-confidence. Louise has also experienced some problems within her friendship group, due in part to her own insecurities.

Louise visited the school many times before her transition to help her find her way around the school site. Louise regularly sees the Visually Impaired Service who monitor her inclusion and advise school of any equipment which Louise could make use of. Louise currently uses a CCTV magnifier, Lunar ICT Enlarger software and a hand-held magnifier to assist her in class.

Louise's normal seat is at the front in all geography lessons, with the light coming into the room behind her. Louise was escorted from one lesson to another at the start of term but now knows her way around. Louise has support in practical subjects from teaching assistants for safety reasons alone; she has no support in geography.

Louise's class have been learning about the different ways that rocks are formed and the uses people have for different rocks in her recent geography lessons. Louise has brought her completed homework to the lesson; she has annotated an enlarged diagram of the rock cycle which checked her understanding of the formation of metamorphic rocks.

The objectives for this lesson were as follows below.
Students will:

- *Reinforce their knowledge of the three types of rocks – metamorphic, sedimentary and igneous.*
- *Learn how to use a classification system by sorting relevant from irrelevant information.*
- *Be able to describe what different rocks look like.*

Louise will:

- *Complete the same activities as the other students without a visual impairment.*
- *Share her views with others to boost her confidence.*
- *Realise that magnifying technology aids us all, not just those with an impairment.*
- *Use other senses besides sight to classify the rocks.*

Additional support

- *The teacher gave all instructions from the board, well away from the window, which distorts Louise's vision.*
- *All paper resources were enlarged twice the size – A5 to A4, A4 to A3.*
- *The teacher incorporated the CCTV enlarger into the lesson by using it to model how to examine the rock samples. Louise and her partner then both used her CCTV enlarger to view the rocks.*
- *The questions in the envelopes were printed on orange card to reduce the glare for everyone.*
- *The questions asked all students to touch the rocks, describe how they felt etc, which played to Louise's strengths.*

VISUAL IMPAIRMENT PROFILE	
Potential impact of characteristics on learning	• May require a differentiation in terms of curriculum contents, access, resources and outcome. • May require additional curriculum, i.e. specialist teaching of Braille, mobility, tactile, keyboard, social and life skills. • May be unable to access the curriculum via the visual media required. • Take longer to process and assimilate information and require additional time to complete tasks and activities. • Students need additional time for their eyes to adapt to impulses from the environment.

	• May require additional format for reading and writing, and access to specialised resources, e.g. Braille and writer. • May become visually fatigued when concentrating for sustained periods of time. • Limited perception of environment. • Poor spatial awareness.
Possible supportive teaching and learning strategies	The type of specialised teaching strategies and resources required will differ from student to student. Expert advice and training must be sort from a specialised teacher. Ask the student to identify his/her requirements. Additional strategies may include some or all of the following: • Ensure optimum position for student in terms of impairment, light source, inclusion etc. • Person teaching should not position themselves against the light. • Implementation of an extended curriculum and specialised teaching strategies and resources such as the use of Braille, tactile maps, keyboards, appropriate print and diagram size. • Restructuring of the environment to enable the student safe access to all areas, materials and resources, e.g. space to move, replacement of reflective surfaces, appropriate lighting, position in class in relation to light source and access to power source etc. • Structure environment, clearly identify if changes have taken place and walk student through the new layout if necessary, and ensure floor is free of clutter. • Use a multi-sensory approach, whenever possible. Use images, sounds, tastes and smells. • Allow additional time for students to observe and assimilate what they perceive. • If appropriate, adapt material and resource in terms of size, clarity and contrast of print against background. • Ensure equipment and resources are available in a different format, i.e. large print, tape, Braille, electronic documents, tape for class and examinations if required. • Allow time for student to relax following periods of intense concentration. • Provide only information required, avoiding clutter and unnecessary details. • Provide oral, simple drawings or written descriptions to support photographs etc. • Use appropriate correct coloured paper and type print that meets the visual needs of individual students. • Whenever possible supplement learning by using objects, artefacts, tactile displays, pictures, graphs, models. • Produce tape recorded descriptions of regularly used maps and photos.

	• Liaise with the local university geography department for use of raised simplified line maps. • Use of models for landforms.
Preparation and planning for outdoor learning	• Discuss with students, parents and specialist teachers for the VI to identify possible issues and solutions including any additional staff support needed if appropriate. • Pre-visit essential. • Hand-held tape recorder.

Royal National Institute for the Blind (RNIB) 105 Judd Street, London WC1H 9NE Tel: 020 7388 1266 Fax: 020 7388 2034 Website: http://www.rnib.org.uk

Hearing Impairment (HI)

Hearing impairment is a broad term that is applied to partial or complete loss of the ability to hear. The loss may vary in both nature and severity. Hearing impairment may occur through a conductive loss which is usually temporary and ranges from mild to moderate or sensorineural deafness which is permanent and varies in degree. Students who have partial hearing may use hearing aids which amplify sounds but may also distort them. People with this degree of hearing impairment may still be able to communicate effectively orally. Students whose degree of loss is complete may use a variety of alternative communication language forms such as British Sign Language. Depending on the severity of the condition a hearing impairment may affect the language, social and emotional development of a student. It may also impact on academic achievement. Students with a hearing impairment may cover the whole ability range.

Main characteristics:

These will vary from student to student depending on the effect of their specific hearing impairment but may involve the following:

- use a variety of communication methods
- good visual skills
- difficulties in interactional and social skills
- emotional difficulties and poor self-esteem
- isolated and withdrawn
- become tired and fatigued after periods of intense concentration

Paul

Paul is 12 years old and attends his local high school. He has a sensorineural hearing impairment and wears hearing aids. He uses speech but this is often stilted and unclear especially when he is angry or excited. He uses BSL to support communication. He is a happy and highly motivated young man with a 'wicked' sense of humour. Although there have been difficulties in the past he has worked hard at developing his social and interactional skills and now has a close but small group of friends.

He sits at the front and to the left in class with the window behind him. This position enables him to make best use of his hearing on the right, supports his ability to lip read, access written information, and provides him with a clear view of his teacher.

HEARING IMPAIRMENT PROFILE	
Potential impact of characteristics on learning	• May have difficulties with language structure and literacy. • May miss important information. • May find it difficult to participate in or follow group discussions. • Requires more time to assimilate and respond to information, questions etc. • Will become fatigued when lip reading as this requires intense and sustained concentration. • May require an interpreter or note taker. • Environment needs to be light and as free from background noise as possible.

Possible supportive teaching and learning strategies	The type of specialised teaching strategies and resources required will differ from student to student. Expert advice and training must be sort from a specialised teacher. Ask the student to identify his/her requirements. Additional strategies may include some or all of the following:
	• To support communication:
	  • Obtain the student's attention before you speak.
	  • Speak clearly and at a normal pace, do not shout.
	  • Support the student's ability to lip read by not covering your mouth with your hand, pen etc, putting your back towards the student, standing so the light is behind you and your face is in shadow.
	  • Do not talk while writing on the board.
	  • Rephrase or repeat information if required.
	  • Provide additional information in written or visual form.
	• If the students are using an interpreter or note taker always speak directly to the student.
	• In group discussions sit students in a circle and identify a set of rules and cue to manage the interactions, e.g. each person in turn makes a contribution.
	• Clearly outline lesson focus, activities and expected learning outcomes and support in writing.
	• If possible provide information prior to lessons.
	• Supplement oral or signed communications with written or visual material.

	• Use short sentences, they are easier and quicker to comprehend than long ones. • When possible begin explanations with concrete examples before moving on to more abstract ones. • Present only one source of visual information at a time. • Maximise the use of visual media. • At every opportunity obtain feedback from the students to determine understanding.
Preparation and planning for outdoor learning	• Provide step-by-step information regarding organisation, requirements, focus of study, activities and expected outcomes prior to visit. • Provide peer or group support to take notes or written text – it is impossible to write and lip read at the same time.

Royal Institute for the Deaf (RNID), 19–23 Featherstone Street, London EC1Y 8SL Tel: 0808 808 0123 British Deaf Association (BDA) 1–3 Worship Street, London EC2A 2AB British Association of Teachers of the Deaf (BATOD), The Orchard, Leven, North Humberside HU17 5QA www.batod.org.uk

Multi-Sensory Impairment (MSI)

The term multi-sensory is used to represent a diverse group of students who may have varying degrees of visual and hearing impairment which may be combined with other disabilities. The complexity of their needs may make it difficult to ascertain a true picture of their intellectual abilities. The impact of dual sensory impairment may have a significant impact on the student's ability to gain information, meaning and understanding from and about their environment and may result in sensory deprivation. Teaching approaches must facilitate the effective use of all the senses including residual hearing and vision within an environment that ensures the development of trusting relationships and a consistency of approach.

Main characteristics:

These will vary considerably from student to student but may include some of the following:

- visual and hearing impairment
- additional disabilities
- idiosyncratic methods of communication

- challenging or self-injurious behaviour
- isolated and withdrawn behaviour

Lizzie

Lizzie has a dual sensory impairment and a physical disability. She wears hearing aids and uses a working frame to support her mobility. She has some useful peripheral vision that she uses effectively with touch to explore her environment. Lizzie is non-verbal but uses a range of strategies including vocalisation and gross motor movements to communicate her emotions, likes and dislikes, choices and responds her environment. The staff use personal indicators, routines, objects of reference and enlarged black and white bold outlined key symbols to support meaning, communication and comprehension. The classroom is structured into clearly defined areas to facilitate the use of her residual vision and provide her with the security and confidence to explore and gain meaning from her environment.

Lizzie demonstrates that she is aware of patterns that form part of her daily routines. Within the classroom and wider internal and external school environment she can locate differentiated areas and associate familiar sensory stimuli, objects, activities, events and people to these areas. She enjoys accessing environments beyond the school and in particular enjoys the sensation of soft winds and the smell associated with wooded areas, and responds positively when revisiting these experiences through simulated sensory experiences.

Lizzie occasionally presents challenging and self-injurious behaviour when she is confused or frustrated. She is working at P 3(ii) in all areas of the National Curriculum. Her teacher has an additional qualification in the teaching of students with dual sensory impairment.

MULTI-SENSORY IMPAIRMENT PROFILE	
Possible impact of characteristics on learning	• Restricted information through vision and hearing. • Access to incomplete or distorted perceptions of the world around them as a result of sensory limitations. • Inability to sequence and link activities. • Aversion to contact with aspects of their environment including people. • Difficulties in establishing trust and developing relationships. • Difficulties in accessing, making sense of and processing information. • Does not pick up information incidentally. • Becomes tired and fatigued after periods of intense concentration.
Potential supportive teaching and learning strategies	Physical environment • Calm atmosphere. • Clear physical structure which provides safe pathways for mobility. • Identify different areas and details of the environment and link these with regular activities.

	• Use of matt surfaces to avoid glare. • Effective and appropriate lighting levels. • Environment and working areas free from clutter. • Furnishings used to control noise levels and distractions. • Blinds and curtains used to control glare from windows. Development of Relationships • Each member of staff has a badge coloured or with a symbol, to identify who they are to the student. The student is alerted to adult's presence and to when adult leaves. • Consistent approaches provided through routines and responses of staff. • Give students opportunities to control and act upon their environment and make choices. • Allow the student time to respond. Communication • It is supportive if all of the student's communicative behaviour is responded to consistently. • Appropriate modes and aids to communication should be identified and implemented. • Frequent opportunities should be provided for communication to take place. • Commentaries provided to inform the student what is going to happen, what will happen next and when activity is finished. Teaching strategies • Learning should be broken down into small steps but taught as part of the whole activity. • Allow time for the student to respond. • Concrete examples should be used. • Revisit experience through simulated sensory activities.
Preparation and planning for outdoor learning	• Pre-visit location to ensure access and facilitates appropriate. • If appropriate rehearse visit in familiar surroundings or through simulated sensory experience.

MENCAP, 117–123 Golden Lane, London, EC1Y 0RT Tel: 020 7454 0454
Website: http://www.mencap.org.uk

Physical Disability (PD)

There is a wide range of physical disabilities and students with PD cover all academic abilities. Some students are able to access the curriculum and learn effectively without additional educational provision. They have a disability but do not have a special educational need. For other students, the impact on their education may be severe, and the school will need to make adjustments to enable them to access the curriculum.

Some students with a physical disability may need to miss lessons to attend physiotherapy or medical appointments. They are also likely to become very tired as they expend greater effort to complete everyday tasks. Schools will need to be flexible and sensitive to individual student needs.

Cerebral Palsy (CP)

Cerebral Palsy is a persistent disorder of movement and posture. It is caused by damage or lack of development to part of the brain before or during birth or in early childhood. Problems vary from slight clumsiness to more severe lack of control or movements. Students with CP may also have learning difficulties. They may use a wheelchair or other mobility aid.

Main characteristics:

- *spasticity* – disordered control of movement associated with stiffened muscles
- *athetosis* – frequent involuntary movements
- *ataxia* – an unsteady gait with balance difficulties and poor spatial awareness

Students may also have communication difficulties.

Gus

Gus is a bright, enthusiastic student who has an opinion on everything and gives 100% to all he does at school. Gus has cerebral palsy which makes him forgetful, a little disorganised at times and tired at the end of the day. Gus loses concentration as the day progresses and tires more quickly as the term progresses. Gus walks with an unsteady gait and struggles to control pens, pencils and rulers etc. as his fine motor skills are poor. Gus has to take his time to speak and some students in his class can get irritated by this. Gus uses a laptop to type the majority of his school work and leaves lessons a few minutes early to get to his next lesson before the rest of the school fill the corridors.

Gus receives support from all five of the school's teaching assistants. He has a Statement of Special Educational Needs. The aims of the Statement are to ensure safety in school, to further develop his fine motor skills, to improve his spoken language, and to make progress in all basic skills to have full access to the National Curriculum. These aims are to be achieved with any appropriate facilities or equipment, and any modifications to the National Curriculum. The school liaises with speech therapists and occupational therapists as necessary. An IEP has been prepared by the school which details short-term targets for Gus and useful classroom strategies.

Background – rainforests

In the first lesson of the unit the class discussed and fed back what they already knew and understood about the rainforest. Gus, as always, had some prior knowledge and told the class about sloths and how they move slowly through the trees.

Strategies for Gus's inclusion

- *Preparing the room before the class arrived so Gus would not have to move furniture*
- *Ensuring that one of the groups was next to a plug for Gus's laptop*
- *Planning a lesson which drew on memories. Gus can think faster than he can articulate his ideas orally*
- *Allowing Gus's teaching assistant to act as a scribe for the brainstorming, but giving Gus the responsibility of producing the final piece of work*
- *Setting the homework early in the lesson when Gus is always present*
- *Devising plenaries that check understanding without relying on a lot of student explanation*
- *Allowing Gus to leave early without missing the plenary*

PHYSICAL DISABILITY PROFILE	
Possible impact of characteristics on learning	The impact on learning will vary in relation to the degree of the disability, any associated medical conditions and/or the degree of learning difficulty but may include some of the following: • Difficulties with mobility, movement, maintaining posture, fine motor co-ordination and skills. • Poor attendance due to illness or medical and therapeutic appointments. • Poor attention, concentration, organisational and memory skills. • Easily distracted. • Spatial or positional perception problems. • Co-ordination of visual motor skills. • Speech and communication difficulties. • Poor basic skills. • Difficulties in the physical process of writing. • Poor confidence and self-esteem. • Fatigue.
Possible supportive teaching and learning strategies	• If appropriate seek advice of other professionals, i.e. physiotherapist, occupational and/or speech therapist. • Structure environment by ensuring that appropriate resources are available to provide a good physical seating and working position.

	• Investigate the need for additional resources to support physical aspect of writing and alternative approaches to recording. • Build on prior knowledge, reinforce abstract concepts with supplementary visual resources and concrete materials. • Reinforce new skills and generalisation by using a range of methods in different settings and repetition. • Check understanding through questions. • Support work using material from the board by highlighting key information. The student may find a tape recorder helpful or an adult may scribe if appropriate. • Allow additional time to meet physical and personal needs and to complete tasks and activities. • Address communication directly to student not through the teaching assistant.
Preparation and planning for outdoor learning	• Discuss with student, parents, school nurse, any possible issues that may arise. Produce care plan to address specific physical or medical needs. • Pre-visit location to ensure access and facilitates appropriate. • Staff pre-read handouts and outline tasks to be accomplished before field work. • Provide adapted recording sheets or alternative forms of recording. • Allow extra time for completion of tasks. • Organise working to form part of paired or group activity and differentiation tasks within members of the group.

Scope, PO Box 883, Milton Keynes, MK12 5NY Tel: 0808 800 3333 (Freephone helpline) Fax: 01908 32105 Email: cphelpline@scope.org.uk
Website: http://www.scope.org.uk

Tourette's Syndrome

Tourette's Syndrome is a neurological disorder which is characterised by 'tics'. Tics may be simple or complex and consist of involuntary, rapid or sudden movements and uncontrollable vocalisation which is repeated over and over again. Rarely vocalisation may include inappropriate words or phrases. Students may sometimes suppress their tics for a short time, but the effort is similar to that of trying to control a sneeze until the tension becomes so great that the tics 'explode'. Finally the pressure mounts to the point where the tic escapes. Tics may increase in frequency and severity in exciting or stressful situations and decrease when a student is relaxed or engrossed in an activity.

Main characteristics:

Physical tics

These range from simple blinking or nodding through more complex movements to more extreme conditions such as echopraxia (imitating actions seen) or copropraxia (repeatedly making obscene gestures).

Vocal tics

Vocal tics may be as simple as throat clearing or coughing but can progress to be as extreme as echolalia (the repetition of what was last heard) or coprolalia (the repetition of obscene words).

Tourette's Syndrome itself causes no behavioural or educational problems but other associated disorders such as Attention Deficit Hyperactivity Disorder (see pages 49 to 51) or Obsessive Compulsive Disorder (OCD) may be present.

Michael's Story

I was diagnosed with Tourette's Syndrome when I was 12 years old. Before that time people thought that I was naughty, rude and hyperactive. Many people still do. My tics included facial twitches, shouting out and making funny noises. Sometimes the tics are really bad and then my legs jump, my head moves from side to side, my arms swing out and I also shout out. This is very hard for me and I try to control the movements by tensing my body or concentrating hard on doing different things but this does not always work. Sometimes I can hold the tics back but then it is like I am bursting; the tics explode and go on for a longer time than usual and then I disrupt the class. My school and teachers have been very helpful but I have been bullied. I am sometimes very angry and frustrated and hit out at other students and teachers. Sometimes I also feel very down in the dumps. I have only a few close friends which I can understand as my behaviour embarrasses me so it must embarrass others.

TOURETTE'S SYNDROME PROFILE	
Potential impact of characteristics on learning	• Disruption of tics on student's attention, performance and achievement. • Impact of tics on social relations and self-esteem. • Verbal abilities may be better developed than others that rely on the manipulation of visual information. • Difficulties organising work, memory and copying. • Repetitive behaviours. • Copying information quickly and accurately from the board can be particularly difficult. • Additional behavioural issues.
Possible supportive teaching and learning strategies	• Accept the student, the tics are not deliberate. • Provide a tolerant, compassionate classroom ethos. • If student is in agreement discuss the impact of Tourette's Syndrome with the class.

	• Provide personally challenging but a stress free teaching and learning environment with the facilities to access a private area for study or when taking examinations if required. • Allow additional time for student to complete work or examinations. • Allow student to sit where he/she feels most comfortable. • To lower stress give instruction in stages. • Provide worksheets, tape recorders or computers to support reading and writing. • Arrange a shared signal that the student can use to show that they need to leave the room as tics are becoming overwhelming, and a private place for them to go.
Preparation and planning for outdoor learning	• Explain expectations and tasks well in advance. • Provide friendship group working or 'buddy' system.

Tourette's Syndrome (UK) Association PO Box 26149, Dunfermline, KY12 7YU
Tel: 0845 458 1252 (Helpline) Tel: 01383 629600 (Admin) Fax: 01383 629609
Email: enquiries@tas.org.uk Website: http://www.tas.org.uk

Summary

It is hoped that the above outlines will be helpful to you in a number of ways. You may use them for your own professional development, to heighten your awareness. You may refer to them when planning and preparing curriculum activities in geography. They may be useful to share in departmental meetings to use as a stimulus for discussion about how to break down the barriers to learning. They are not stereotypes, but informed summaries of a variety of conditions. These profiles will help you to support each individual student with respect and dignity. By sharing this expertise we hope that you will be able to create learning in geography that is accessible, relevant, worthwhile and, above all, enjoyable for both you and your students.

References

1. Special Educational Needs Code of Practice (2001) London: DfES/581/2001
2. QCA (2001) *Planning, Teaching and Assessing the Curriculum for Pupils with Learning Difficulties: Geography*. London: QCA www.nc.uk.net/ld/Ge_content.html

CHAPTER 4

Creating the Inclusive Classroom

Introduction

> The purpose of education for all children is the same, the goals are the same but the help that individual children may need in progressing towards them may be different[1].

This is a timeless and challenging position to take on education. It is a perspective that must pervade a school. Most schools provide a broad range of educational opportunities through a set of curriculum aims and objectives. This curriculum informs and helps students make decisions for themselves about their own lives. Schools actively teach subject specific curricula but they are also one of the main places where students learn about the hidden curriculum. Beliefs and attitudes are challenged, shaped or reinforced by schools. In Appendix 4.1 Bev Rowley outlines how geography contributes to these wider school curriculum aims at Stretton Brook School (also on the CD).

Of course all these aspirations have significance for the way that we organise the classroom, the geography of our learning space if you like. Here again the subject matter of geography itself has a keyrole to play. You might like to spend a moment or two, considering Rob Kitchn's[2] observations on geography and disability.

> In the case of disability, geographers are interested in how landscapes reflect the values of the people who live there. Geographers are also interested in how such designs and values affect disabled people – how disabled people feel when they are trying to negotiate inaccessible environments and how this affects their spatial behaviour (where they go). Such studies demonstrate that disabled people's spatial behaviour is not just affected by issues of accessibility

but also by issues of acceptance, provision and attitudes. Understanding how geography disables people, then, is as much about understanding how the environment conveys messages of belonging and exclusion as it is about understanding the organisation and structure of places.

How does this help you to think about your classroom, and secondly, how does it help you to think about how places are represented through your geography curriculum choices?

Why is an inclusive classroom important?

The implementation of the Disability Discrimination Act 1995[3] should now ensure that public and private organisations adequately provide for and make arrangements to include people with disabilities. Schools likewise are increasingly modifying buildings and purchasing specialist resources to include those with some sensory, physical or cognitive difficulties in addition to providing for those with learning difficulties and emotional and behavioural difficulties.

This is because many believe that mainstream provision:

- facilites social inclusion and is able to break down negative stereotypes
- is able to offer a wide-ranging curriculum

For some students, special schools will provide the most appropriate setting. These schools have much to offer, not only to their students but also in terms of sharing their experiences and expertise with mainstream colleagues.

Sooner or later your head teacher will need to ensure that all teachers have the necessary knowledge, skills and understanding of disability and disability issues to provide a broad and balanced curriculum suitably differentiated to take account of students with a wider range of disabilities.

As many schools develop their physical site for inclusion, many will have access to a sensory room. This has huge potential for geography. 'Using a Sensory Room to Develop a Sense of Space', in the appendices and on the CD, gives some ideas for using this facility. Some of these ideas can equally be used in a usual classroom environment.

How can geography exclude?

If we agree that geography is about people's experience of places and their ability to recognise their own place in the world then it is easy to see how geography can exclude.

Exclusion takes many forms. Buildings exclude people physically by preventing access. Imagine how your sense of place would suffer being in a wheelchair when there are only steps to the entrance. Try to imagine reading small signs in a public building pointing you to where you want to go if you are visually impaired.

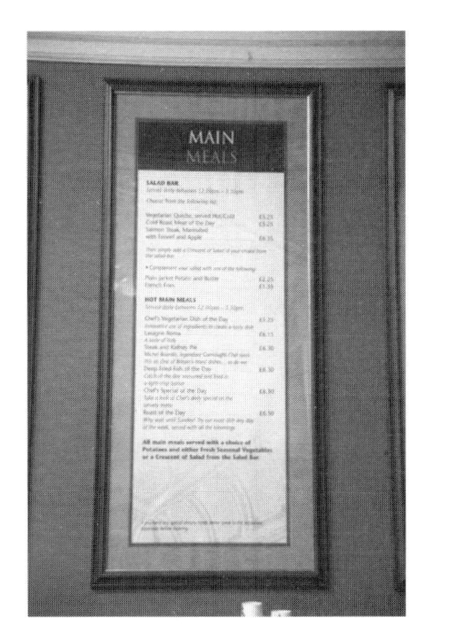

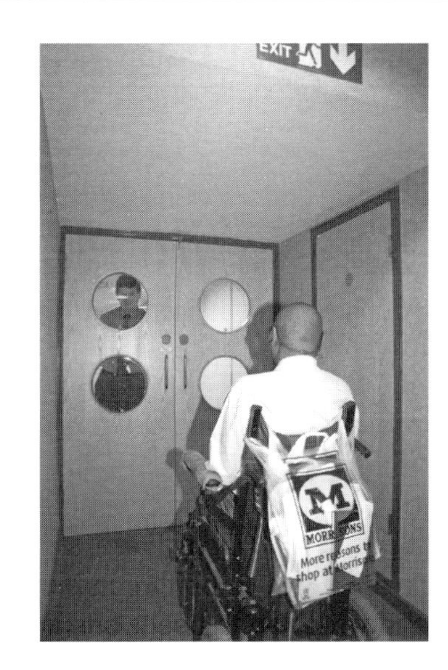

Society can segregate and exclude people in a variety of subtle ways. As well as physical exclusions there is social exclusion, and exclusion based on misinformation and poor understanding. How would you feel if your able-bodied friend had the choice of holiday destinations but you could only choose from one or two? How far would that change your perception of the places you visited? Would all landscapes fill you with the same awe and wonder?

Take a moment to think about how a busy shopping centre/restaurant may provide sensory overload to people with ASDs. How would you cope if walking from one end of the street to the other took you twice as long and proved to be a stressful, dangerous, disorientating experience?

How many times would you negotiate the uneven kerbs? Would you keep trying or give up? Would you retain your enthusiasm for discovering new places? How long would it take for the town centre to become a disabled-free zone if people did give up? What messages would that then convey about public places and society's values? Exclusion changes behaviours through limited choices and opportunities.

Think Global . . . . . Act Local . . . . . modifying classrooms to aid the national agenda

If we agree that in studying geography we appreciate the decisions people make and how these affect an environment and the people who live there, we should appreciate the importance of the decisions we make in our classrooms.

To include students in geography lessons we need to ensure the following:

- that equal opportunity is given to each student to experience and take part in the lesson

- that we promote independent learning by supporting learning appropriately
- that we accommodate freedom of movement and ensure our lessons are user friendly

In the world outside the classroom many subtle changes are being made to the built environment to aid inclusion. Steps and ramps are now being fitted to buildings, door handles are being fitted at lower levels, and door widths are being widened to accommodate wheelchairs. In offices and public buildings audio-taped copies of letters and policies are being made available for the hearing impaired. Street furniture is increasingly contrasted and more obvious to the eye, and tactile maps and guides are being produced for those with visual impairments.

Schools and classrooms can easily be audited, and recommendations made to modify sites to accommodate people with physical or sensory disabilities. Local Education Authorities are working through schools to make necessary modifications. Local modifications throughout society will change global practice and attitudes.

Resources are needed to buy specialist equipment, but there are many things that teachers can do which do not cost money or require too much time. In fact it is often the subtle changes to the craft of teaching that make all the difference.

Each school will have advisory teachers linked to their school for students with sensory or physical disabilities and the range of learning difficulties. Advisory teachers inform schools of the needs individual students have. Each student is different and the needs of two students with visual difficulties will not necessarily be the same.

There are some generalisations to be aware of but each teacher should liaise with the school's SEN department to ensure they have up-to-date information about all students and the specific strategies to use with particular students.

Towards a more inclusive indoor classroom

Geography teachers try to help students possess a detailed understanding of how places are different. We teach about people, their choices and opportunities, the decisions that are made that govern their lives and communities; the climate and weather, the soils and geology, ecosystems, changing land use, tectonic activity . . . The list goes on . . ..

In dealing with the subject content, literacy and numeracy skills will invariably be needed. Many students on a secondary school's SEN register will have general difficulties with literacy and numeracy.

An inclusive indoor classroom for all

All students learn best when they are at ease, relaxed and confident to ask for help. They need to know that both the teacher and the other students will support them to learn a new concept, and encourage them to keep trying when they get it wrong.

- Everyone should listen respectfully when anyone else is speaking.
- Volunteering or offering answers should be positively rewarded and incorrect answers sensitively dealt with.
- Students should be encouraged to care about their own right to learn.

Develop a relationship with the students in your room.

- Don't assume students know nothing about the controversial issues in geography.
- Be available and approachable for discussions, show that you want to listen.
- When questioned, be prepared to share appropriately your own views in a reasoned and rational dialogue. You may also find it helpful to admit when you don't know the solution to a geographical issue.
- Re-assure students when new and difficult tasks or activities are set.
- Expect socially unacceptable opinions; see how they change through the unit.

Seating arrangements can help support students.

- Students with attention deficit problems are sometimes best sat well away from windows or other distracting areas of the room.
- A students with cerebral palsy who uses a laptop may need to be sat near a power point.
- Think carefully whether it is appropriate to ask all students to move furniture. Should you then ensure you move it before the lesson begins?

Classroom space can be maximised to help support learning.

- Word walls displaying topic words and new terminology help students recall language.
- Colour coding and labelling equipment help students to work independently.
- Display boards showing students' work, if they are changed regularly, help students recognise their achievements.
- Visual resources stimulate the majority of students but can overload those with sensory impairments such as Autism.

Using the whiteboard, interactive whiteboard or blackboard.

- Use the board for reminders but not for large pieces of work.
- If the board must be used give students a photocopied transcript.
- Using different colours at the beginning of each line or numbers at the start and end of each line will help students keep their place.
- Writing on the board should be larger than you would normally write, clear and well-spaced.
- Allow plenty of time for students to read from the board, and check their work after they have finished.
- Handouts should always supplement any work that is on the board. When in a mixed ability setting try to sit dyslexic students and visually impaired students facing the board and not have them sat at an angle.

Preparing and using resources

- Check the readability level of your worksheet against that of the student – keep the writing style in short, simple sentences and use short paragraphs.
- Check the typeface and size of print – is it clear and large enough? Use fonts such as Ariel or Comic Sans, keeping the font size 12 or more.
- Use bold to highlight and avoid underlining titles as it makes words run together.
- Check the quality of copy – avoid blurred images.
- Check the spacing – avoid clutter. Try using boxes between paragraphs to break up text. Wider spacing between sentences and paragraphs should be used.
- Background graphics can make text difficult to read. Photocopy on to coloured paper instead of white.
- Provide line guides to help students keep their place and provide a dyslexic dictionary to encourage independence. Equip students with a set of highlighters to help with notes.

Below is a list of resources to consider having to hand in all classrooms:

- dyslexic dictionaries: contact the Dyslexia Institute for current advice
- dyslexic spelling logs: contact the Dyslexia Institute for current advice
- phonic dictionaries
- traditional dictionaries/thesaurus

- subject specific spelling lists
- word mats/walls – for each topic and skill
- line guides for handwriting
- line trackers for reading
- non-slip matting
- coloured dyslexic overlays
- highlighter pens
- colour coding stickers
- electronic spell checkers
- simple calculator for general numeracy
- light pencils and soft rubbers for mistakes
- number squares
- range of coloured board pens for teacher to use
- laminated systematic crib sheets for subject specific skills

Giving clear instructions

- Instructions should be specific and kept short and simple. Students with SEN should only be given one, maybe two instructions at a time.
- Gain eye contact with students who struggle to pay attention and for whom this is not inappropriate to their special needs. It may be helpful to the student if the teacher or TA repeats the instructions to them quietly.
- Ensure that there can be no ambiguity read into what you have said . . . is it optional or not?
- Reinforce instructions regularly, go back and ask students to tell you what they have to do.
- Provide visual prompts to help students remember what they have to do.
- Repeat and re-use key vocabulary so students learn how to use it appropriately.
- Keep maps and diagrams as clear as possible.

Incorporating the technology

Many classrooms now have TVs, videos, stand alone machines, networked computers, laptops wired to digital projectors, interactive whiteboards, CD-ROMS and software packages that promote independent research and learning. These can be useful in geography because:

- physical processes can be brought to life through animated sequences
- distant places can be brought to the classroom
- viewpoints and opinions can be heard and empathy can be developed with characters
- images and music can be linked to provide added stimulus, building a more complete picture of places, helping auditory and visual learners connect with the work being covered
- editing, cutting and pasting can help teachers create interesting resources.

Using maps

Maps, like all resources, will need adapting according to the needs of different students, but here are a few points that might help your thinking. Maps are a key form of visual communication used by geographers. They are used as sources of information and as a means to represent information spatially. Much can be done to make maps accessible to many students.

- Use, alongside the map, a handout with the area heavily simplified and using thick lines.
- Use a school-based Geographical Information System to recreate maps with only key information highlighted, or to facilitate the use of alternative colours that are more appropriate to the student's needs.
- Reduce the information complexity, think about the learning objectives, why you are using the map, what the specific information is that is needed by the student.
- Emphasis the line work on the map.
- Sonic and tactile maps may be accessible by linking up with a local university geography department.
- Think about making tape recorded descriptions of the maps commonly used in your curriculum.
- Think about creating digital images to go alongside regularly used maps.
- Think about creating a sensory box relevant to particular landscapes represented on a map.
- Think about supplementary text to accompany particular places on maps.
- Photographs related to specific map locations will also be of benefit.

- Whenever possible avoid using colour as the only way of conveying information, use it as a supplementary medium. For example, use the labels 'sea' and 'land' in addition to the colours on a map of a coastal area.
- Ensure that text and graphics stand out against their backgrounds.

Celebrating achievement

- Use credit marks, merit points, stamps or stickers to reward good work.
- Reward effort as well as content.
- Reward pieces of work that are deserving of the praise, do not patronise students by going overboard.
- Write positive comments in exercise books and use the students' names to personalise achievements.
- Regularly discuss the students' positive progress with them.
- Discuss each student's IEP with them and make time to mention when IEP targets have been met.
- Consider setting up a department policy to encourage students: for example, geographer of the week, sending praise letters home.

Promoting independence through re-usable models

Students need strategies to help them approach new material. Explaining suitable methodology helps, but students need to be able to apply that methodology and transfer the approach to new material to be independent learners.

Below is a list of models that can be re-used, repeated and re-visited to promote independence.

1. Locating places on a map
   (Use informal grid references: the UK is a witch on a pig – is the Holderness coast on her hat or the pig's foot?)
2. Drawing mental maps
   (Encourage students to locate a place, imagine standing there, visualise the places, work out where places and features are in relation to each other, fill in the gaps)
3. Promoting a sense of place
   (Ask students questions about what they would see, hear, touch, smell and taste in this new place)

4 Describing landforms and landscapes
(Encourage students to look for colours, shapes, textures and patterns in landscapes)

5 Using the Development Compass Rose (see CD and below)
(Teach students to ask questions about the Natural, Economic, Social and Who decides (Political) environments to find out about a place)

6 Reinforcing the physical and human difference
(All environments are different, structure your approach to investigating the difference. Physical differences are in the climate, relief, rocks, soils and vegetation. Human differences are social, political, economic and cultural)

7 Using enquiry questions
(Ask the same questions for different case-studies. Ask What? Where? Who? Why? So What? What could? What Should?

8 Using writing frames
(Create stand alone re-usable writing frames to help students with such areas as presenting an argument, writing persuasively or writing for an audience) 'How do I write better descriptions' provides a helpful example, see page 86 and the CD.

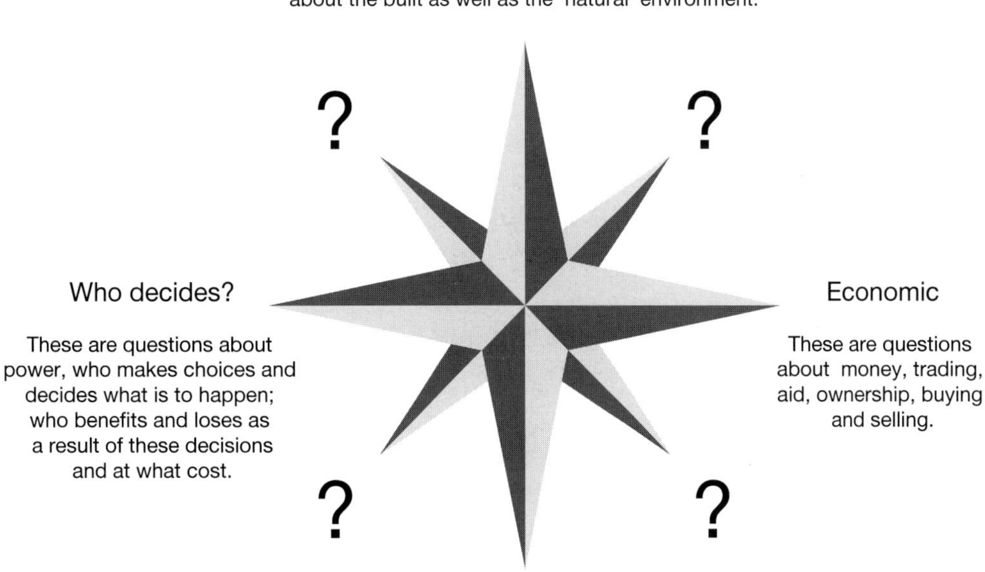

Source: *The Development Compass Rose*. Development Education Centre, Birmingham, 1995[4]

Reducing the barriers to learning in geography

The geography department should feel a responsibility to help all students improve their standards in literacy and numeracy. Schemes of work should identify the specifics of geography's contribution to literacy and numeracy. For some students it will be their geography lessons that inspire them to achieve their highest literacy and numeracy levels. This will be because they find the content particularly motivating or because the learning style demanded is one that they particularly excel at.

Strategies could usefully be developed to help reduce barriers to learning in geography. For example:

- possessing a lack of geographical imagination
- poor possession of language to describe places
- lower abilities in recalling and applying geographical enquiry to writing structures
- poor ability to imagine places from maps
- difficulties with number calculations
- inability to estimate
- difficulties in extracting meaning from graphs
- poor understanding statistics and ratios
- lower ability to recall shapes and locations
- poor ability to interpret patterns

There are activities and exercises that can easily be incorporated into geography lessons to contribute to the development of understanding in geography by reducing these barriers to learning.

Using literacy and numeracy to support extended writing in geography

A large number of Staffordshire teachers are now using a series of models to help students improve three key areas of their geographical writing. There are three modes which are used as scaffolds to help students see how to write with more depth. More information can be found by visiting www.sln.org.uk/geography/BPRS[5].

Teachers know what a student understands by how they respond and contribute creatively and orally in lessons. Inevitably there will be some form of written assessment at some point. The models provide a re-usable guide to help students see where they are and what they have to do to improve their answers.

Difficulty	Activity
Language to describe places	• Stepped development from words to sentences, use of adjectives to build from • Encourage the use of models to help work out descriptions from a photograph – spot colours, shapes, patterns etc. • The use of word mats may also be helpful here (see appendix 5.3 and CD resources for chapter 5).
Poor imagination of places	• Read accounts of places from various sources • Use multi-sensory resources
Making distinctions between different types of writing	• Exercises to determine the difference between what is true/false, fact/opinion, theory/fact • Turn sentences from fact to opinion etc.
Difficulties in recalling and applying geographical enquiry to writing structures	• Use writing frames to promote the organisation and structure of ideas (descriptive, explanation, discussion, persuasive) • Use images to support writing, include story boards
Abstracting meaning from graphs	• Living graphs, matching events in a story to those on the graph • Regular practice of reading off graphs, including activities where students calculate, interpret and extrapolate
Understanding statistics and ratios	• Matching diagrams to numbers • Calculating • Pictograms
Ability to recall shapes and locations	• Informal place tests • Using shape of countries to help locate features • Use clear outlines • Use appropriate colours • Maps from memory
Imagining places from maps	• Link photos with maps • Read text descriptions of places • Use taped descriptions • Empathise, virtual fieldtrips
Interpret patterns	• Display specialist vocabulary in a variety of visual ways. For example, nucleated, scattered, clustered, linear, adjacent, meandering

By focusing on developing these key geographical components of writing, teachers can provide students with a security blanket, they can use the models until they can write fuller answers independently. Scaffolding techniques help all learners but provide students with learning difficulties with the structured support they need.

Writing better descriptions

Students are regularly asked to describe landscapes, climates, census data etc. in geography lessons. The model encourages students to describe people and places in an increasingly more complex way. The model promotes recognising that there are simple extremes or opposites (hot, cold) in phenomena. There are also different types of that phenomena (less economically developed countries, more economically developed countries, newly industrialised countries) and elements of it can be compared and should be compared by trying to find definite relationships between the data (twice as much, 70% of, one third of). Students will write more complex descriptions when they can recognise ratios and patterns within phenomena.

For an example lesson using this frame, see 'A lesson plan using the descriptions scaffold' on the accompanying CD. You may also wish to refer to 'Writing about development: using writing frames in geography' on the CD.

HOW DO I WRITE BETTER DESCRIPTIONS?

How do I do it?		What do I write?
Extremes	Use place names to identify where something is. Using extremes or opposites to tell us about something.	Newcastle Under Lyme . . . Staffordshire . . . Large/Small . . . Wet/Dry . . . Busy/Quiet . . . Full/Empty . . .
Different types	Recognise there are variations or categories or different types in what you are studying.	High growth/steady growth/no growth . . . The Tundra/the Tropical Rainforest/the Desert. The elderly people/the young people with families/the teenagers/the disabled.
Comparisons	Use numbers to compare features.	Twice as many people . . . half the number of visitors . . . a third less money . . . calculate the average, the range, percentages from data.
Ratios and Patterns	Spot different types, use numbers and group these together to tell us about the whole place. Try to find a pattern or relationship.	As the temperature increases the rainfall decreases . . . the further away from the town centre you go the fewer big shops there are . . .

Writing better explanations

Giving reasons for the standard of living in shanty towns, the melting of the ice caps or the effects of tourism is confusing. The explanation model encourages students to ask increasingly complex geographical questions to extend their explaining. Students' most basic explanations give one simple reason as an answer. A better explanation links several causes and effects together to explain a phenomenon, or attempts to explain how the causes link together to create an effect. Even more complex answers take this further and recognise how multiple causes work to create multiple effects which then cause other phenomena.

HOW DO I WRITE BETTER EXPLANATIONS?

How do I do it?		What do I write?
Cause Effect	Explain that one feature is caused by another.	A one sentence answer – adding . . . due to . . . because . . . and . . . also . . . as well as . . .
Cause Cause How? Effect Effect Effect	Explain how one feature is caused by another. Or explain how one feature causes another which then has a knock on effect and causes something else.	A group of sentences – cause and effect. . . . this is caused by . . . and so this means that . . . this affects this by . . . the consequence is . . .
Cause How? How? Effect	Explain how one feature causes another and then how the knock on effect caused something else.	A paragraph – sequencing . . . this is the result of . . . the consequence of this is . . . this means that . . . next . . . then . . . meanwhile . . . finally . . . after
Causes Effect Effect	Explain how two separate causes **work together** to create one feature which then causes something else.	Two paragraphs, one about each cause – emphasising . . . the combined effect is . . . this leads to . . . these two things then create . . . this happens when . . . whereas . . . above all . . . significantly

For an example lesson using this frame see the Year 11 Flood Control lesson plan, on the accompanying CD (A lesson plan using the explanations scaffold).

Writing better judgements

Students are regularly asked to form an opinion about subjects as diverse as deforestation, the building of a by-pass or the Kyoto Treaty. We hope that the opinion the student possesses is informed. We try to encourage students to base that opinion on either facts, evaluation of evidence or empathy for the people involved. The judgement model encourages students to increasingly look for evidence and weigh up the importance of individual and collective pieces of information. (See accompanying CD for the 'Writing better judgements in geography' resource.)

A note about mixed ability teaching

Think for a moment about how you would respond to this teacher.

> I am an NQT in geography who trained in a school with set classes. My new school has mixed ability groups. How can I ensure that every student's learning needs are catered for, in particular those with SEN? Differentiating tasks tends to mean that planning my lessons can take a very long time. Is there any advice on how to effectively plan and teach lessons to the benefit of all students in a mixed ability teaching group?[6]

In this and the previous chapter there is a range of specific advice to help, develop, reform and refine your practice. However there are also some more general teaching tips that could usefully form part of the answer to such a question.

Be realistic and think about the mixture of abilities that you will find in each class. Students have all sorts of strengths which can be exploited to benefit one another. A student with a visual impairment may have poor spelling and be slow to produce written work but may have a fantastic sense of place. All students can learn from each other.

- Where possible try to supplement reading and writing with the use of visual and/or sensory stimulus. Clear photos, effective maps and diagrams, the use of artefacts for students to hold, plants, smells relating to places, all provide useful stimulus material.
- Where appropriate offer students a choice of presentation styles. Some like to use ICT, whilst others like to represent ideas pictorially.
- Encourage all students to highlight any specialist vocabulary that they are uncertain about to help them be active participants in the lesson. The interactive whiteboard may be helpful here.
- Encourage students to use and create appropriate maps as often as possible. This helps them to think spatially about places and to develop a confidence about mapping as a geographical tool.

Summary

One thing that we sometimes overlook in the classroom is that some students simply just need extra time to explore, create, question and experience as they learn (see the 'Writing about development' resource on the CD). By revisiting established geography curriculum plans or by creating new learning activities with a view to reducing the barriers to learning, all the students in our classes stand to reap the rewards of effective and appropriate teaching and learning.

References

1. Warnock Report. Department of Education and Science (1978) *Special Educational Needs: Report of the Committee of Inquiry on the Education of Handicapped Children and Young people.* London: HMSO
2. Kitchin, R (2000) *Disability, Space and Society.* Sheffield: Geographical Association
3. Disability Discrimination Act (1995) London: HMSO
4. Development Compass Rose, www.tidec.org.uk
5. Staffordshire Best Practice Research Scholarships, www.sln.org.uk/geography/BPRS
6. Adapted from query on inclusion website http://inclusion.ngfl.gov.uk

CHAPTER 5

The Outdoor Classroom

Introduction

> A school which is striving to provide well for the children in most need will be raising standards for all.
> The Code of Practice, 2001[1]

This chapter is concerned specifically in supporting planned activities that take place outside of the classroom. This includes fieldwork and residential visits. It also includes any learning that takes place in the local area or school grounds. This is significant for geography.

> Geography without fieldwork is like science without experiments; the 'field' is the geographer's laboratory where young people experience at first hand landscapes, places, people and issues, and where they can learn and practice geographical skills in a real environment. Above all, fieldwork is enjoyable . . . It provides an opportunity for pupils to relate in new ways to each other and to their teachers, and so to break some of the patterns of behaviour that become fixed in a school environment.[2]

We recognise that working with students in the outdoor classroom is both exciting and challenging, motivating and draining. As teachers we experience huge contrasting emotions, often questioning if the energy that we put into organising outdoor learning is worth the benefits. We would strongly argue that it is, and so have constructed what we hope will be a chapter of support. By reading these next few sections we hope that you will develop a confidence about planning valuable, worthwhile and essential outdoor learning. QCA on their *Innovating with Geography* website (www.qca.org.uk/geography) state that:

> Geographical Fieldwork provides opportunities for the first-hand investigation of places, environments and human behaviour. It is a statutory part of geographical education for all pupils at key stages 1–4. Fieldwork grabs pupils' interest and provides a relevant 'real-life' stimulus for geographical questions setting up a sequence of investigation, collecting, recording, presenting, analysing and evaluating evidence as part of geographical enquiry . . . Fieldwork provides opportunities for pupils with preferences for visual-spatial, bodily-kinaesthetic, interpersonal and naturalist learning styles.[3]

When designing fieldwork activities we have it in our gift to integrate accessible learning opportunities, and to remove barriers to outdoor learning. This is both liberating and intimidating. The key is to facilitate access to learning through planning appropriate activities. Many of the principles and guidance will be much the same as for the indoor classroom. Much of the advice and support given in Chapter 4 is also of value when thinking about learning outdoors.

When we consider learning outdoors and special needs it is useful not only to think about appropriate ways to access the geography curriculum, but also what geography as a discipline has to offer. The fieldwork that some students engaged in when they were at school is now informing their university research. For example, specific geographical research on visual impairment may result in practical outcomes that bring positive, practical benefits.

> Another area in which geographers and cartographers are able to provide practical help to disabled people is the development and testing of navigation aids for people with visual impairments, and by providing accessibility maps [see pages 7–8, Chapter 1 in this text]. The aids currently being developed to help people with visual impairments to navigate independently through the built environment can be divided into two groups: in-field and learning-based. The aids include talking signs (audible beacons sited in the environment which visually impaired people follow), personal guidance systems (which link satellite positioning technology with an electronic map to guide a person through an environment by giving directions), and tactile strip maps (maps that can be learnt through touch).[4]

This is also the sort of technology that we can look forward to using on school visits in the future.

Why is learning outdoors so significant?

It is through fieldwork that students develop a sense of place and a sense of awe and wonder about their environment. It is here that they undertake 'real world learning'. The geographical data and information that is gathered is then used to inform and deepen their original reactions to a place. Sometimes these places are being seen for the first time by these students. This is exciting and worthwhile learning for both geography and the wider curriculum. If a student is busy or anxious this will detract from the development of their sense of place.

In the lesson plan included on the CD and in Appendix 5.1 (Local Leisure – Wheelchair Enquiry) Bev Rowley describes a piece of fieldwork considering wheelchair access to local leisure facilities and how this work contributes towards developing students' understanding meaningfully.

Fieldwork encourages students to:

- visit particular areas that illustrate a particular geographical theme or feature
- collect and then present results
- analyse and evaluate the findings of real experience, work out how that then fits in with the theory of other places
- make conclusions and justify results
- create their own their work, and to discover their own interests
- develop their personal and social awareness and an appreciation of others
- develop an appreciation for the environment
- develop a sense of place; fieldwork encourages an emotional response to landscape
- develop a confidence in their ability to be safe outdoors

Fieldwork creates opportunities for teaching and learning.

- many field visits have cross-curricular potential
- often there are opportunities to develop vocationally related skills
- transferring learning from the classroom to the field and back again, and then applying this to other places, helps to strengthen geographical understanding

Learning outdoors and special educational needs

When thinking about fieldwork we may also need to consider the needs of students not usually thought to have special needs, for example students with asthma, diabetes, epilepsy, musculoskeletal disorders, lung and kidney complaints, heart problems and chronic fatigue syndrome. In these cases it is best to enter into a dialogue with the students and their parents and/or carers about the most appropriate steps to be taken to access outdoor learning.

Awareness of who the first aiders in the group are should be shared and communicated, as should a list of any allergies. Information about where the epipen is stored should be known to all, and reference should be made to medical care plans. Remember too if there are any diabetic students with you, there should be a designated member of staff to keep an eye on when and how much they are eating.

There is no single universal difficulty. All students are unique regardless of their capabilities or disabilities. There is therefore no standard template to support special needs students in the outdoor classroom. Each student needs to be considered individually. Often the fieldwork is such a significant learning opportunity that individual discussion and negotiation both with the students themselves and others (including parents and carers) is essential. This will help to facilitate an enjoyable and effective experience. Whilst students with special needs may experience certain challenges they also have strengths. It is helpful to consider ways that they can share these strengths in a group situation outdoors.

When planning appropriate activities it will be beneficial to focus on the geography specific outcomes that you hope the outdoor learning will facilitate. This will then enable you to concentrate on how to reduce the barriers to learning to access these objectives. For example, whilst on fieldwork, to analyse soil samples a physically disabled student may be asked to analyse soil near a flat surface whilst the remainder of the class may scale a hillside. The teacher in this case was careful to ensure that the difficulty in analysing the soil type was the same since this was a piece of GCSE coursework. Such strategies to break down the barriers of learning should be welcomed by all.

Before going on a field trip, or engaging in any outdoor learning, draw up a list of the proposed activities and rate each according to the barriers to learning that it might pose for your special needs students. Use this information to select the activities that are going to be most beneficial to all. For the activities that may be more difficult to access, consider the learning objective. Is there an alternative but equivalent activity that could be constructed?

You may find Appendix 5.2 particularly helpful here (also on the CD). It highlights the main barriers to learning and suggests groups of learners that may gain access to outdoor learning if supportive teaching strategies are used to break down these barriers. The table is not exhaustive, nor should it be used to promote stereotypes. The table used in conjunction with the detailed information in Chapter 3 should help you highlight which outdoor learning activities might benefit from review. It may also help you to select the most appropriate activities to engage in. All students will have a variety of needs, and the combination of these, together with their character, is what helps to make them unique. Emily is a Year 8 student, with a condition known as ataxia, which basically means 'no co-ordination'. The case study on the CD describes Emily in more detail and how provision was made for her to access fieldwork activities and her other learning in geography.

Teresa Lenton, a secondary PGCE course co-ordinator, has constructed a useful mnemonic to help remind us of considerations when planning fieldwork experiences (see www.geography.org.uk/gtip/gtip_4c.html). This also appears on the CD. You may wish to photocopy and annotate it at a departmental meeting. It will be useful to consider a piece of fieldwork that the department currently offers. Use this to annotate the mnemonic with any issues that reduce the quality of experience for students with special needs that you have in an identified group. Now use the table in Appendix 5.2 and Chapter 3 to consider how you would reduce these barriers to learning.

F unding
I CT?
E qual Opportunities
L egislation
D esign
W here? When?
O rganisation
R isk Assessment
K inaesthetic learning

A note about risk assessment

Within your school you will have a member of staff who is designated as the Educational Visits Coordinator (EVO) who will work alongside you to advise on the school and the local authority's guidelines for facilitating outdoor learning safely. The decisions that you take regarding reducing the barriers to learning should form part of your risk assessment. The Field Studies Council can provide knowledge and practice of the content of risk assessments http://www.Field-studies-council.org. On the CD, Julie Dale provides an example of a risk assessment and also illustrates how this has been adapted for a student with special needs (Risk Assessment Example and Modifications to Dovedale fieldtrip).

Teachernet (www.teachernet.gov.uk/visits) provides information on school journeys and outdoor education centres and on health and safety on education visits, as does a poster produced by the Geographical Association (www.geography.org.uk/download/REfieldworkposter.pdf).

The RGS-IBG website (www.rgs.org) provides information about the Expedition Advisory Centre, which offers information, training and advice for anyone embarking on scientific or adventurous expeditions.

The Risk Assessment can be a hugely valuable document for all staff and students if it is a 'live' document. You will find it useful to share this with other teachers and teaching assistant colleagues who accompany the visit. It may also be appropriate to use some or all of it, either in its existing, or in an adapted, format with your students, parents and carers. This will help to clarify expectations and make everyone aware of the strategies to use should the need arise. The more informed fieldwork staff and participants are, the more likely it is that they will have a successful, enjoyable and safe experience.

Pre- and post-visit support

Many departments find that the time invested in creating digital, video, taped and sensory resources related to favourite fieldwork localities are of huge value to many students with varying needs. They of course take time to construct, but over time the benefits will be there for all to experience. Some departments created place boxes. In here they put tape recordings of landscapes and maps. They develop sensory trails and have other materials relating to the place, e.g.

soil, sand and pebbles (small!). Digital images are manipulated and have symbolic communication systems used with them. For an excellent example as to how to facilitate this, visit www.widgit.com/rainforest/ html/start.htm. Here a rainforest locality is developed using photographs and the Widget Rebus Symbols Software (see www.widget.com).

Other departments have collated their various locality resources and created accessible websites that students can use both before and after the visit itself.

If you are planning to create a web-based or visual resource to support pre- and post-visit activities here are a few things to consider. For example you will need to think about how you use colour to convey meaning. This can unwittingly create barriers to learning for colour-blind users. A sharp contrast between background and text can be disorientating and uncomfortable for dyslexic users, whereas insufficient contrasts can make access difficult for those with visual impairments. One solution could be to make the design flexible, so that it is easy to change the colours and text size by making quite simple adjustments.

Students with autism can have unusual fixations on parts of objects. When using a computer this can mean that they are able to focus totally on the screen and to block out all distractions around them. Using a computer before and after the visit with digital images and information pages may help the students to access the field learning in a more secure and comfortable way. They may feel more in control of their learning. You can buy software that can help autistic students with social concepts such as learning to understand facial expressions.

These 'virtual' resources should always been seen as supplementary to the visit itself. Offering a solely virtual fieldwork experience could perhaps be seen as an example of exclusion rather than inclusion. Inclusion is facilitated by modifying practice, adjusting destinations and adjusting learning outcomes. In other words by developing supportive teaching strategies that overcome potential barriers to learning outdoors.

Using the school grounds to support learning outdoors

It is helpful if students are able to have a range of opportunities to support outdoor learning in a variety of locations. Often rehearsing these in the school grounds can be hugely supportive. Outlined below are several ways that the grounds can be used both as part of the geography curriculum and as preparation for fieldwork visits off-site:

Using a compass
Developing trails
Treasure hunts
Using your senses to investigate . . .
Following directions
Investigating weather, where are the hottest, shadiest places etc.
Decision making, where should we put . . . litterbins . . . the new school bench etc. . . .
Expressing likes and dislikes
Placing images/representing places

Measuring rainfall
Creating questions, I would like to find out more about . . .
Developing geographical stories, what if . . .?
Developing an awareness of change
Appreciating the familiar.

Overcoming barriers to learning outdoors

As we discussed earlier there is no universal template to facilitate access to outdoor learning. The same is of course true of the indoor classroom. However, you and your department may find the following points helpful to consider. You may wish to enlarge and photocopy them and use them to prioritise how you may adjust your fieldwork activities. (This list is also available on the CD.)

Overcoming barriers to learning outdoors

- All fieldwork details to be given out in advance in a variety of formats, e.g. spoken, written, pictures so that queries can be sorted out prior to the visit.
- Travel – are there any particular issues related to getting to and from a particular site? What about access at the site itself?
- Write down for all students potential emergency situations and how they will be expected to behave.
- Summarise for hearing impaired students any debates/role play activities. Ask all students to raise their hand before they speak so that lip readers can face them in a debate situation.
- Provide written details about the main features to be seen in the field and the activities and projects to be undertaken. This benefits a deaf student and clarifies the learning to be experienced by all the students on the trip.
- Take digital images or video a fieldwork location that is not accessible to a student in a wheelchair. These may also be used with other classes as part of the pre- and post- fieldwork activities.
- If you have students with you who lip read think about where you or any other presenters are standing. Lip reading a person who is standing in front of direct sunlight is not easy!
- Link with a local university department who may have access to inject facilities that create tactile maps. These allow visually impaired students to navigate maps by touch.
- If the visit involves look and see type activities consider using carefully selected digital images to supplement the experience. These can be accompanied by clear line sketches.
- Extra time can be of benefit for all. Do not rush – the temptation to overload the day can reduce rather than enhance learning.

- If you are able to have TAs with you on the visit make certain that they have been appropriately briefed about what to do when and with whom.
- When completing questionnaires think about using a buddy system or facilitating the use of a hand-held recorder.
- Consider distractability in the field. Try not to give instructions in places with lots of background noise, or with lots of activity going on.
- In planning for fieldwork that includes wheelchair users consider:
    - time: it may take longer for the students to get on and off vehicles, or in and out of particular locations
    - access to toilets
    - giving adequate breaks so that students don't become uncomfortable and can change position
    - weather: wheelchair users may need to consider more or different clothes
    - carrying clipboards on a wheelchair needs thinking about!
    - specialist transport may have financial implications, try to make certain that you are aware of this
    - who is to push the wheelchair?

Some of these ideas can then be tailored to a student's specific needs. For example Phoebe is a dyslexic student going on a day's fieldwork to Castleton in the Peak District. In order for Phoebe to access the learning, her teacher and teaching assistant made certain that:

- travel arrangements were clear before they left
- all fieldwork information was provided in a variety of formats, verbal, written and emailed to her home
- Phoebe had several reminders before leaving on the trip
- her teaching assistant spent some time in lessons working with her on time management skills
- Phoebe was allocated to work with a group of students
- she was allowed extra time on the trip to read information
- Phoebe had all the visit handouts before going on the visit
- she was given a list of unfamiliar and key words and place names
- all the handouts were on yellow coloured paper (see advice in Chapter 3)
- back in school the use of a word mat was found to be helpful for key words relating to tourism (see Word Mat on CD).
- many of the handouts were lists of bullet points, with wide spaces between them

Summary

Perhaps the final words should be left with the students. It is they who will persuade you that the adjustments that you make to your outdoor learning activities are well worth the time and energy that this work will undoubtedly demand of you. These comments were posted by students on the CBBC Newsround site during September 2004 (http://news.bbc.co.uk/cbbcnews/).

'If you hide away from life, what will you achieve? Take a risk, take the plunge! Who knows, you might even enjoy yourself! I really enjoy school trips, you might even learn something!!!' Zoe,12, Nottingham.

'I think we should have school trips as they increase learning through interest levels. Its stupid not letting us go on school trips for safety reasons as accidents can happen inside school as well as outside!' Ashlie, 14, Atherton.

'Of course we need school trips because you learn better having a first hand experience. In my school we have field trips all the time and most of the kids do great and it really makes learning that much more interesting' Irene, 13, Canada.

'Yeah I'd much rather we got out of the classroom to experience the real world. We shouldn't be wrapped in cotton wool!' Steve, 13, Darlington.

References

1. Special Educational Needs Code of Practice (2001), London: DfES/581/2001
2. Bland, K., Chambers, B., Donert, K., Thomas, T., (1996) 'Fieldwork' In *Geography Teachers Handbook*, Bailey P and Fox P (Eds). Sheffield: Geographical Association
3. QCA Innovating with Geography Website, www.qca.org.uk/geography
4. Kitchin, R (2000) *Disability, Space and Society*. Sheffield: Geographical Association

Additional references

Inspiration for this chapter came from the Higher Education Research Council funded Geography Discipline Network's website on Fieldwork and Disabilities in University Geography. A useful starting place to access this work is at http://www.rgs.org/category.php?Page=5expedis

Other helpful sites include:

www.rgs.org

www.geography.org.uk

www.teachernet.gov.uk/visits

www.widgit.com/ideas/intro_2_symbols/index.htm

www.fieldfare.org.uk/ Fieldfare works with people with disabilities and countryside managers to improve access to the countryside for everyone.

http://www.fieldfare.org.uk/gpg.htm links to 'A Good Practice Guide to Disabled People's Access to the Countryside'. This publication contains standards and guidelines that will help more people with disabilities to visit and enjoy the countryside.

CHAPTER 6

Monitoring and Assessment

Introduction

Assessment is 'the process of gathering, interpreting, recording and using information about pupils' responses to an educational task'[1]. This statement is helpful, since it supports us in considering how we value students' work in geography and how we use this information to support them to improve their abilities and capabilities in geography. Here it is helpful to look back at Appendix 3, which outlines some of the opportunities and challenges of a geographical education for students with special needs. It is worth also reminding ourselves that students may have been identified as having special educational needs, but that they may also have particular strengths that enable them to achieve in a geography curriculum context.

A helpful summary of the principal features of educational assessment can be found in Appendix 6.1, also on the CD. Assessment should not be seen as a one way process. It is more than a means of collecting data that can be used in end of year reports. Through careful monitoring and assessment of different elements of students' learning in geography, teachers develop an insight into the strengths and weaknesses of students. This informs planning and allows teaching and learning to be adapted to meet the needs of the individual.

Assessment is more effective if students play active roles in identifying their own achievements and areas for development. It should be a dialogue. There is much to be gained by simply talking with students. Following a piece of residential fieldwork, students were asked by their teacher to carry out a self-assessment activity. Here are some responses:

'Fieldwork is the best thing about geography; a good way to learn, but it's difficult for me to make notes in the field, especially in bright sunlight'.

'I enjoy fieldwork but can't walk far and get very tired, so it's hard to work in the evenings.'

'It's difficult to remember all the instructions when you are outside . . . I need regular reminders.'

Feedback such as this, that helps students reflect on how and what they achieved and their barriers to learning, and challenges the teacher to develop further their teaching activities, can only be of benefit. This is an example of how assessment is of value only if it is then acted upon; this is assessment for learning, formative as opposed to summative in its nature.

Summative assessment is the assessment **of** learning. It provides a summary of the level that students have reached in their learning. It can allow us to make judgements about the effectiveness of teaching by providing measurable data about how an individual has progressed in a topic. This can be achieved by assessing the students' learning on the same criteria both pre- and post-topic. This enables progress to be measured.

Formative assessment is often referred to as assessment **for** learning. It supports teaching and learning by providing feedback to both teachers and learners. Formative assessment is used to inform the planning to enable students to identify and achieve their next goal.

A fair system of assessment will allow all students to gain the information needed but in a way that is appropriate for their needs. For example, if a visually impaired student is given longer to analyse rock samples as part of an assessment activity then it is important that the class as a whole appreciates the reason.

Assessment for learning

> Assessment for learning is the process of seeking and interpreting evidence for use by learners and their teachers to decide where the learners are in their learning, where they need to go and how to best get there.[2]

Assessment for learning is guidance that is designed to support teachers in using assessment to raise students' achievement. It is based upon the principle that students will achieve higher standards if they understand the aim of their learning, where they are in relation to this aim and how they can achieve their aim. Appendix 6.2 presents a helpful summary of learning activities that present opportunities to assess students' work (also on CD). The table on page 100 usefully highlights the implications for the work of teachers in the classroom (also included on CD). The review from which this is derived emphasises what teachers in classrooms can do to avoid the negative impact of tests on motivation for learning. It also indicates the actions that can enhance motivation for learning. To accomplish these goals teachers should:

do more of this . . .	and do less of this . . .
Provide choice, when appropriate, and help students to take as much responsibility as they can for their learning	Define the curriculum in terms of what is in the tests to the detriment of what is not tested
Discuss, using appropriate communication strategies, with students the purpose of learning and provide feedback that will help the learning process	Give frequent drill practice for test taking
Encourage students to communicate in an appropriate medium how much they have learnt and what progress they feel that they have made	Teach how to answer specific test questions
Develop students' understanding of the goals of their work using appropriate strategies and provide feedback to the students	Allow students to judge their work in terms of scores or grades
Help students to understand where they are in relation to the learning goals and how to make further progress	Allow test anxiety to impair some students' performance
Give feedback that enables students to know the next steps and how to succeed in taking them	Use tests and assessments to tell students where they are in relation to others
Encourage students to value effort and a wide range of attainments by using different and appropriate strategies	Give feedback relating to students' capabilities, implying a fixed view of each student's potential
Encourage collaboration among students by using appropriate groups/pairings; this will help to develop a positive view of each others' attainments	Compare students' grades, giving status on the basis of test achievement only
	Emphasise competition for marks or grades among students

Adapted from p.8, Assessment Reform Group, 2002[3]

Systems for assessing and recording students' attainment

Assessment is an essential part of reflecting on learning. It should be an essential part of curriculum planning. This includes enquiry based learning in geography. We have found the 'framework for learning through geographical enquiry'[4] extremely helpful here to place assessment in its curriculum context. The figure on the following page illustrates the framework. Planning, assessing and recording are essential elements that underpin this framework (included on CD).

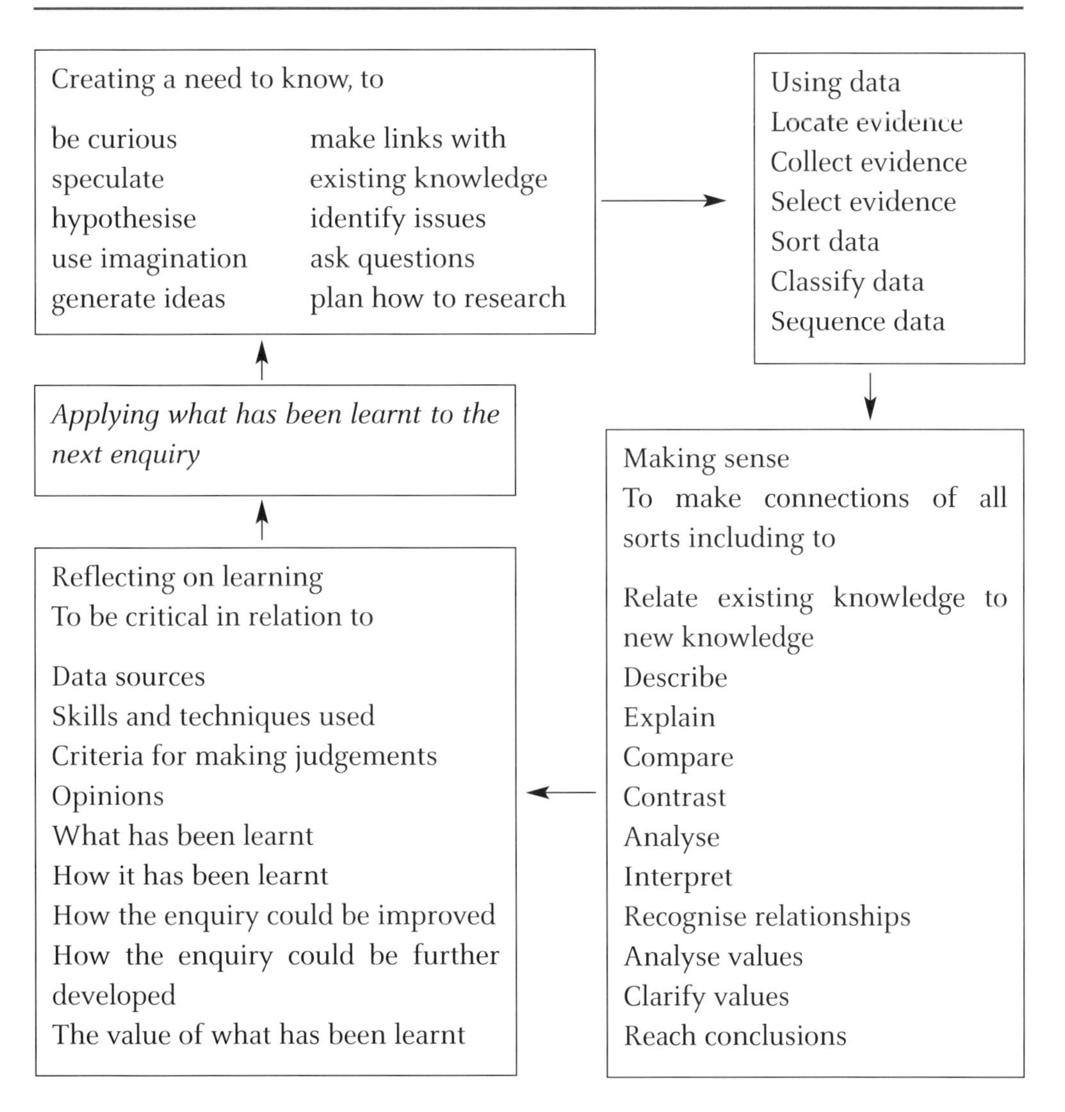

Planning should be based on positive achievement. It should recognise the vital role that assessment plays in planning future work and in enabling students to progress. Learning objectives need to be clear and achievable, stating precisely what the teacher wants the students to learn. The learning activities are based upon the learning objectives. These are the small steps which build upon the knowledge, understanding and skills within a unit, promoting progress in students' learning. Assessment opportunities, both formal and informal, should be an integral part of planning. This will enable the teacher to make a judgement about what the students have actually understood, their acquisition of knowledge and what they are able to do. Some activities may provide particularly good opportunities for assessment. The Earthquake poster on the CD illustrates one way that a piece of work can be assessed and how the assessment can be used to inform planning. Here, Bev Rowley shares some of her work from Stretton Brook School. It illustrates a moderated analysis that was made of a student's work. The photograph was used to support and inform the teacher's judgement – it was not used exclusively and this is significant. Assessment often requires us to make rounded and informed judgements.

The geography programmes of study can be modified for students with learning difficulties. Material from earlier key stages can be used. Existing knowledge needs consolidating and reinforcing as well as the introduction of new knowledge, skills and understanding. Staff can choose to focus on particular aspects of the programmes of study in depth. First hand experience (or virtual experiences) are important to support students in gaining both knowledge and understanding. This may be developed through sensory activities, contact with different people or fieldwork.

Planning needs to be monitored and evaluated as an ongoing process. Schemes of work may need additions, such as extension activities or the modification of activities. Annotations can highlight aspects in need of development. Teachers may find some learning activities more valuable and decide that extra time can be allocated to improve understanding.

Assessing allows professional judgements to be made about what the students know, understand and can do.

Teachers need to employ a range of assessment strategies to assess learning and inform teaching:

- discussion
- observation of students working
- teachers' notes or log
- feedback through marking or grading work
- in-depth written comments
- tests
- student assessment
- peer assessment
- compiling a portfolio of assessed work
- pre- and post-unit assessments

Through the constructive use of such assessment strategies, the setting of targets for improvement in geography will enable students to understand what they need to know to progress further. These targets for improvement should be used to inform planning and maximise opportunities for achievement. Some judgements are informally noted by the teacher, however, other areas of progress or concern need to be recorded.

Recording provides a formal record of progress.

Departments, and ultimately schools, need to formulate an effective assessment policy to ensure consistency in recording attainment. This will enable staff to understand the students' progress in learning and to decide on the next steps. The system should be manageable, and as far as possible integrated into normal classroom routine. Support staff should be aware of systems in the classroom and, as a minimum requirement, annotate work to show the level of support the student has received. A simple system that students can also use is to write a simple code at the end of a piece of work or lesson, NS meaning no support, SS some support and HS showing a high level of support.

A variety of recording techniques can be used, as follows.

Weekly records

- Marking (adhering to agreed policy, in many schools this involves target marking)
- Annotate students' work
- Teachers notes (particularly useful during fieldwork activities)
- Evaluating and annotating short and medium term planning
- In-depth, formative comments on students' work
- Comments on IEP or geography targets, if addressed
- Photographic evidence
- Video or audio tapes

Termly records

- Evaluate and annotate medium term plans (schemes of work)
- Class record
- Pre- and post-unit assessment marks
- Annotated samples of students' work
- Student awards
- Progress against IEP targets

Yearly records

- Statutory reviews of statements of educational needs
- Progress files
- Reports
- Records of any school tests or external accreditation
- Portfolio of moderated evidence

An excellent example of assessment and monitoring and feedback is illustrated by material featured on the CD. The 'Earthquake poster' was produced by students following some work on the tectonic process; the 'Assessment and Feedback on the Earthquake poster' illustrates the teacher analysis and moderation of this work.

Pre- and post-unit assessments

Pre-unit assessments are useful for ascertaining the knowledge that students have of a topic before it has been taught. A short assessment can be devised for each unit of work to assess existing knowledge. This can take the form of an oral assessment, but is more effective if each student completes a written task. Multi-choice questions are useful for students who may experience literacy problems, as these can be read through as a group and each student can then select their preferred answer. For students with more severe learning difficulties, the assessments can include pictures or symbols. It also ensures that the students have an attempt at all of the questions rather than just giving up and writing 'don't know'.

An example of a pre- and post-unit assessment for some work on tropical rain forests is shown on the CD. Students were asked to complete the same activity at the start and at the end of the unit of work to assess some of the progress that they had made.

Also on the CD, is an example of a pre- and post-unit assessment for some work on settlement.

Post-unit assessments are used at the end of a unit to identify what the student has learnt. It is more effective if the same assessment is given as was used for the pre-unit assessment. This means that progress can be clearly measured. The teacher can analyse the results to identify which students have progressed further and try to identify reasons for this. For example, the unit of work may appeal to particular learning styles or be closely related to other subjects. Alternatively a student may have been absent or experiencing personal problems. The post-unit assessment will provide evidence to prove that an individual target has been met.

Helpful hints

The following points must be considered to ensure that the purpose of the assessment policy is to improve the students' learning. Records such as these will also be helpful to share with others. There should be little or no need to produce extra information for outside agencies such as OFSTED if you are meeting the needs of your students appropriately.

- Teachers can become more aware of their students' learning needs, through observation, joint target setting and Statement of Educational Needs and IEPs. Discussion with others, including learning support assistants, other teachers, SENCO, parents and educational psychologists, can be valuable for understanding the students' learning needs.
- Try to assess the quality of learning rather than the quantity and presentation of work.
- It is often helpful to focus comments on how to improve learning rather than social or managerial matters.
- Try not to overuse empty praise that lacks a learning focus.
- It is always to the students' benefit if approaches are shared between teachers in the same school.

P levels

P levels are used mainly in special schools and primary schools to refine the target setting and assessment process for students working towards National

Curriculum level 1. These performance descriptions outline early learning and attainment in eight levels, from P1 to P8. They can be used in the same way as the National Curriculum level descriptions, to make 'best fit' judgements about a student's performance over a period of time. The performance descriptions can be used to inform short, medium and long term planning and to track progress working towards level 1 of the National Curriculum. They can also be used as part of the target setting process.

Performance descriptors for P1 to P3 outline the general performance that is demonstrated by students with learning difficulties and are general across the subjects. Levels P4 to P8 describe the way in which students' skills, knowledge and understanding emerge in geography. Both the P levels and an example of the P Levels in student-friendly language can be found on the CD and in Appendix 6.3.

Target setting

Target setting has the greatest impact when it focuses on precise curriculum objectives for individuals and when it forms part of the whole-school improvement process.
(Ofsted, 2004[5])

The setting of targets has several benefits for both individual students and for developing effective teaching and learning in geography. On an individual level targets are used to raise the standards of attainment for students. This can be achieved first by gathering detailed information about students' prior attainment.

This can be ascertained from a number of sources:

- the students' statement of educational needs and the annual reviews of these documents
- relevant standardised test data
- individual education plans
- information from other professionals, e.g. educational psychologists
- teacher assessments

The most effective targets are based on assessment information that is accurate. Targets need to be challenging but achievable, and are measurable to enable progress to be clearly tracked over time. They also need to identify and address the individual learning needs of the student. The involvement of students in setting and reviewing targets has proved to be highly effective. It enables students to clarify the steps they need to achieve in order to succeed.

Student progress data, using National Curriculum levels of attainment or P levels, can be analysed to identify both strengths and priority areas for development. This can inform target setting at all levels, including departmental and whole-school targets.

Individual Education Plans

Assessment of the work of students with special educational needs should relate to targets set in individual education plans (IEPs).

IEPs are an integral part of tracking students' progress and informing planning. Students need to have ownership of their IEPs and as with all target setting they prove to be more effective if the learner has been involved in the process. The student and teacher need to work together to identify their strengths and weaknesses. In geography a challenging but realistic objective can then be set for the term, which can be agreed upon.

Staff collaboration will ensure that targets are realistic and measurable across the core subjects. Through termly reviews, IEPs track students' progress. If the target has been achieved then new targets can be set. However, sometimes students need to continue to work on the target. To ensure that the student perceives that progress is being made, the teacher will need to break the objective down into smaller, more achievable steps and devise other strategies to build upon and consolidate the student's skills and knowledge.

IEP targets should be addressed in all curriculum lessons where appropriate, and need to be accessible to teaching staff, support staff, students and parents. The IEPs could be stuck into the students' geography books or folders to ensure they are truly a working document. This needs careful thinking about, as it is important for students with special educational needs not to be singled out amongst their peers. Alternatively a discreet copy of the student's IEP can be kept by the teacher and learning support assistant to ensure targets are being met and progress recorded. Whichever strategy is adopted it is significant that targets are focused on in geography lessons, and that differentiation is taken into account both during the planning and delivery of the curriculum. Any progress can be recorded on the IEP and fed back to the form tutor when the targets are reviewed. Business-type cards can be made by the students to remind themselves of what their IEP targets are. The target needs to be put into language that the student can understand. These can be discreetly carried from lesson to lesson, and taken home.

MY TARGETS
Lauren George S6 (Year10) Summer Term
English
To write simple sentences and short pieces of work.
Maths
To know 2, 5 and 10 times table.
PSHCE
To talk to people I know about things I am interested in.
Geography
To know the compass points, North, South, East and West, and how to use them.

Though these targets are not all specific to geography, they are skills that may be carried across and practised during geography lessons and in many areas of the curriculum. Lauren addressed and made progress against her English target several times during geography lessons, through exercises such as writing three short sentences to describe reasons why Bangladesh floods. Her Maths target has been addressed when she has needed to read and interpret scales on thermometers and rain gauges. Quite often this involves students counting in multiples of 2 or 5 to determine the numbers on a scale. Her PSHCE target of talking to one or more people in familiar settings is addressed during each geography lesson. Her geography target was supported by using a large map of Bangladesh drawn onto a white bed-sheet. (The teacher had constructed this by pinning the sheet to the wall and using an OHT to project a simple outline map on to the sheet, suitably enlarged. The teacher used a marker pen to trace the country outline. The students could then 'walk around' Bangladesh and use the compass points to find and talk about different areas.) With much encouragement, Lauren's confidence in communicating effectively is slowly increasing. However, she still needs prompts when communicating with adults and is reluctant to interact with unfamiliar people.

Instruments used in target setting

A variety of instruments are available for assessing learning, performance monitoring and effective target setting.

PIVATS (Performance Indicators for Value Added Target Setting) is appropriate for students with specific needs that are different from those of other learners. It uses the P-scales and National Curriculum levels up to 4. PIVATS supports inclusion by focusing on the attainment of the student within the stated range rather than on their special needs or age. It promotes value added measurements and promotes school self evaluation and improvement. It also develops individual, cohort, subject and year group profiles of performance which can be used for comparative purposes.

Further information about PIVATS can be found at http://www.lancashire.gov.uk/education/advisory/index.shtml.

Value-added measures

Through setting and measuring achievement against targets, it is possible to track students' progress as well as their achievement. Headteachers, governing bodies and LEAs can then monitor student progress and analyse the extent to which student achievement has been effectively raised. They can look for trends, such as new initiatives, and the extent to which they have affected student progress. This will also prompt discussion amongst staff by identifying areas of strength and areas for development in teaching and learning.

A further purpose of measuring attainment is to identify if specific groups are progressing at a different rate from others, for example are boys achieving more than girls, and are there disparities between different ethnic groups?

Target setting enables the senior management to focus on the school or subject development plan and ensure that any developments support students in the classroom. Teaching and learning should be the central focus. Identifying which elements encourage success and then building upon them is an important step in school improvement and in supporting progress for all students.

Accreditation

A wide variety of external accreditation is available to students with special educational needs. Many focus upon a range of coursework, including written tasks, oral assessments, visual displays and fieldwork. These forms of assessment make the courses more accessible for many students with different learning styles and needs.

When students are entered for external accreditation, close liaison should be facilitated between the student, their parents/carers, their teaching team, the SENCO and examinations officer in the school. An early dialogue will make it so much easier for the student to access the support that will help to reduce the barriers to their success. For example, the request for additional time needs to be appropriately evidenced and submitted early in the Spring term of the examination year. This means that provision needs to be negotiated at the start of Year 11. Often earlier preparation is beneficial. Many requests have to be externally verified, and it can take time to find a suitably qualified person with capacity. If a student is to be entitled to a reader, this can be unnerving, particularly if the student has no prior experience to draw on. It would be beneficial to all if the student could have the opportunity to access a reader during all examination type situations during the course. This would help to take away some of the concern for both the student and others in an already demanding examination situation.

If the accreditation involves a formal examination, this should not be seen as daunting to students as many methods of support are available. Examination boards can be approached prior to the examination to request support. For example, enlarged papers are available for visually impaired candidates. A range of other arrangements can be used, including:

- extra time
- use of a reader or communicator
- use of an amanuensis or scribe
- timed breaks
- use of word processors
- use of practical assistants

The SENCO will liaise with the schools' examinations officer, the student and parents to decide upon the most appropriate support for individual candidates. They will also be aware of any formal applications which need to be made for such support.

GCSE

Since September 2004 forty-five schools are involved in teaching a new Pilot Geography GCSE.

The Pilot GCSE comprises of two parts:

- The Short Course – this is the geography core which is built around the concepts of uneven development, interdependence, futures, sustainability and globalisation. Three themes run through this core.
  Certification is achieved by 67% external assessment and 33% internal assessment.
- The Optional Units – these units develop and support the Short Course. Two units need to be studied from the choice offered.
  The optional units are internally assessed through a range of alternative media ,which enables students to work to their strengths. For example, oral or visual presentations allow greater flexibility and are less restrictive than traditional assessment methods.

Trials of the Short Course of the Pilot GCSE have found that the specifications are more accessible to students in 'vocational' groups and for students with more challenging behavioural needs. This is largely because a greater range of learning activities result in students being more actively engaged. There is also a focus on conceptual learning. Alternative assessment media allow students to demonstrate their understanding in a range of styles. For more information visit www.geography.org.uk/projects/gcse_index.asp.

Entry level certificates

Entry level courses are both inclusive and accessible. The flexibility of these courses allows the needs of the individual to be met and for each student to reach their full learning potential. Entry Level Awards are appropriate for students who at the start of Year 10 are considered to be below the standards required to achieve a GCSE and who are working at National Curriculum Level 3 or below. However they can allow students a grounding that will allow them potentially to progress to GCSE or other vocational programmes.

Geography

Entry Level Geography is intended to be taught and learned in an active way. Fieldwork, practical and investigative work is integral to this, along with units that are relevant, stimulating and accessible.

For example, in the WJEC Entry Level Specification a range of assessment techniques are used including:

- Folios (2 × 20% of final marks)
  Each of these comprises three teacher set pieces of work that have been approved by a consultative moderator.
- Oral Test (20% of final mark)
  This provides candidates who have problems with written work an opportunity to demonstrate qualities, skills and abilities that may not be revealed in written form.
- Fieldwork Investigation (20% of final mark)
  This enquiry should be short and focused. It requires that students define a problem or answer a question about their local area. They will collect and analyse data. Data will be represented as maps, graphs, diagrams etc. and conclusions will be reached.
- Research Topic (20% of final mark)
  The choice of topic to be investigated should be based on the candidate's interests and the teacher's estimate of strengths and weaknesses. ICT should be used during research and presentation. Presentation can take a range of formats such as brochures, newspaper reports, posters, video presentations etc.

Different aspects of geographical learning will be assessed through these components including:

- **knowledge** of places, environments and patterns at a range of scales from local to global
- **understanding** of the course content
- application of **knowledge** and understanding in both physical and human contexts
- use a range of techniques and **skills**

Humanities

This syllabus contains elements of geography (20%), history (20%) and religious education (20%). These core units are assessed by either written or oral tasks which are board set, teacher marked and board moderated.

The Contemporary Issues Unit (20%) will be assessed through three tasks:

- a written test to assess knowledge and understanding
- the production of a visual display to assess skills
- an oral assessment to allow candidates to evaluate evidence

Three of the units relate closely to geography:

- travel, tourism and leisure
- trade and development
- the changing world of work

The Local Issues Unit (20%) is cross-disciplinary. It is assessed through a teacher set task. This is a local cross-curricular research task which can be broken down into three stages:

- a written log outlining the nature of the issue, the student's procedures and the reason for those procedures
- the gathering of information/data/personal interviews etc
- an evaluation of findings. This can be visual display, oral assessment or written report

The assessment objective for the geographical element of this course requires students to demonstrate their:

- **knowledge** – by recalling specific facts relating to the course content
- **understanding** – by describing and offering explanations of spatial patterns and relationships – showing an understanding of the complex inter-relationships between people's activities and their environment
- **skills** – by achieving competence in and making use of a variety of skills and techniques appropriate to geographical enquiry

Through a range of teaching and learning styles this course will allow students to be actively involved in the learning process and achieve positive results, thus enhancing self-esteem.

For more information about these Entry Level Courses contact Welsh Joint Education Committee, 245 Western Avenue, Cardiff, CF5 2YX (Tel: 029 2026 5000; Website: www.wjec.co.uk; Email: info@wjec.co.uk)

ASDAN

The ASDAN Award Programmes allow students to progress through a range of levels. Each programme is designed to develop, assess and accredit key skills and recognise the personal achievements of students. Students are asked to complete a range of challenges. Some modules relate particularly well to geography, especially the Environment module which builds upon a range of geographical skills and knowledge.

The ASDAN Award Programme is often used to motivate disaffected students as it is accessible, flexible and practical. The challenges provide targets for students that are achievable and the students are also involved in action planning and reviewing tasks.

For more information about ASDAN visit www.asdan.co.uk

Summary – assessment and curriculum creativity

Good assessment should promote, not stifle, learning in geography. This philosophy was the basis of some QCA supported work on creativity in geography. An interpretation of creativity (derived from NFER, 1998[6]) was offered by Eleanor Rawling and John Westaway in their Teaching Geography article 'Exploring Creativity'[7]. (Full article and diagram on the CD.)

They stated that:

> creativity is not just another new initiative. It is actively promoted by the National Curriculum and it provides a direct and unifying way of addressing many current demands on secondary schools, such as thinking skills, literacy, entrepreneurial skills and citizenship.

At Stretton Brook school a unit of work on tropical rainforests was developed involving a trip to Birmingham Botanical Gardens. More detail about the visit and the work it generated can be found on the CD ('Creating a Rainforest Environment').

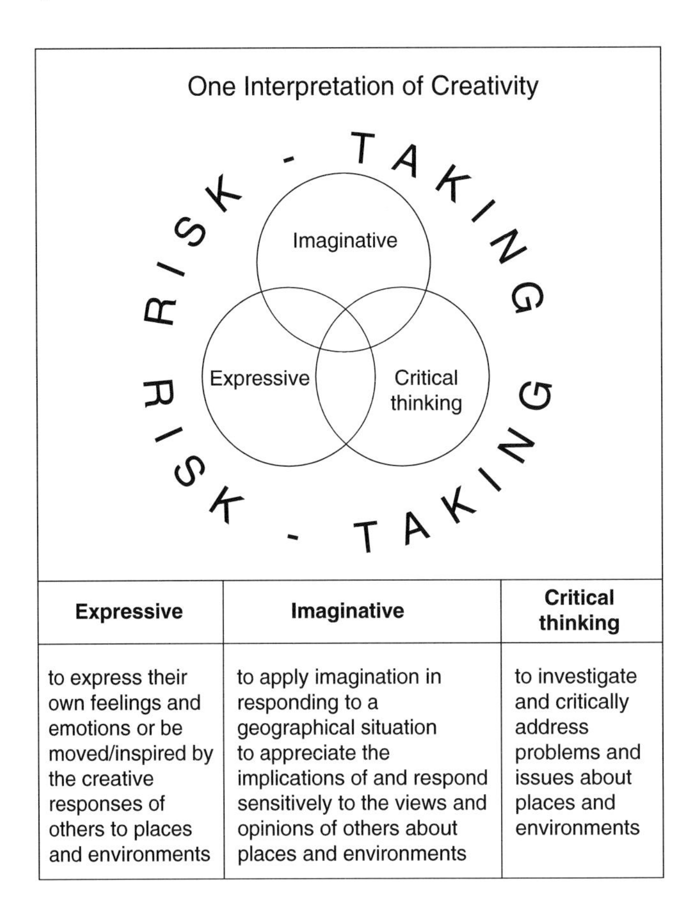

Expressive	Imaginative	Critical thinking
to express their own feelings and emotions or be moved/inspired by the creative responses of others to places and environments	to apply imagination in responding to a geographical situation to appreciate the implications of and respond sensitively to the views and opinions of others about places and environments	to investigate and critically address problems and issues about places and environments

Derived from NFER, 1998[6]

Bev Rowley comments:

> Promoting students' creativity in geography offers an inclusive approach because, where different learning styles are emphasised, more students can be actively involved. In particular, students are able to express themselves in ways other than writing and it is through their imaginative responses that they strengthen their geographical knowledge and skills.[8]

References

1. Harlen, W, Gipps, C, Broadfoot, P, and Nuttall, D (1992) Assessment and the Improvement of Education, *The Curriculum Journal* 3 (3), 215–230
2. Black, P, Harrison, C, Lee, C, Marshall, B, Wiliam, D, (2002) *Working Inside the Black Box, Assessment for Learning in the Classroom.* London; Kings College
3. Assessment Reform Group (2002) *Testing, Motivation and Learning.* Cambridge: University of Cambridge Faculty of Education
4. Roberts, M, (2003) *Learning through Enquiry.* Sheffield: Geographical Association
5. OFSTED, (2004) Setting Targets for Pupils with Special Educational Needs. London: Ofsted publications, HMI 751
6. NFER (1998) CAPE-UK Project Stage 1 evaluation report, www.capeuk.org
7. Rawling, E and Westaway, J, (2003) 'Exploring Creativity', *Teaching Geography* 28 (1), 5–8
8. Rowley, B (2003) 'Creating a Rainforest Environment', *Teaching Geography* 28(1)

CHAPTER 7

Managing the Teaching and Learning Support Team

What is support?

When considering how to reduce the barriers to learning in geography, remember that you are not alone! There are a number of ways that your curriculum can be enhanced. Support in geography classrooms may take many forms, there are the people and there are resources, such as televisions, videos, music, paper resources and ICT that can also support students in their learning. This chapter will concentrate on managing the breathing, walking, talking type of support – the additional bodies that you may have the opportunity to work alongside.

> Support – carry, prop up, give strength to, encourage, endure, tolerate, supply with necessaries, lend assistance or countenance to, back up.
>
> Source: Readers Digest Pocket Dictionary, 1969.

It is a simple straightforward word, yet the appropriate deployment of support underpins success or failure in many a classroom. Classroom teachers of geography should be supported by their Head of Department, who in turn needs to be supported by senior management. Students should be supported by appropriate geography schemes of work, that break down the barriers to their learning.

Supporting students is clearly not just about curriculum coverage. It is about developing strategies to access geographical content in a stimulating way, it is about developing enquiring minds, supporting students to ask geographical questions and investigate their own world thoughtfully and appropriately.

Clearly, teachers need to develop an appropriate relationship with all students in their classes. Brimming with enthusiasm for geography is one thing but without mutual respect, tolerance and understanding little learning will take place. This is even more important with students with special needs, as it is often the small, everyday interventions that makes the relationship a secure and successful one.

So there we have it – support, very much necessary for learning, is about respect and understanding, but most of all is about relationships.

What is management?

Teachers manage, Head teachers manage, and Sixth Form prefects manage. Definitions aside, it certainly encompasses communicating, supporting (that word again) directing, observing, reinforcing and feeding back.

That should be straightforward then . . . we communicate what we want students to do in every lesson, we support and direct them with resources, we observe them working, reinforcing ideas through schemes of work and feedback when there is something amiss or something worthy of praise.

A certain amount of managing is overseeing; assuming what is being overseen is planned properly then there should be no worries there. Of course there are times when plans fail, the unforeseen occurs and chaos ensues, but that's teaching, and likely to occur in geography lessons because we DO practice what we preach about enquiry learning. In short we have all the necessary skills to successfully manage the support we get in classrooms, we just need to unite the team and apply these skills thoughtfully.

Who makes up the team?

There's you, the students and potentially a whole squad of support. Support can be provided in geography departments in many different ways.

Teaching Assistants

Teaching assistants are usually employed by the school to work either with one or two specific students for a few hours a week, or may be employed by the school full time to work across many different year groups or subjects as appropriate with whole classes. Teaching assistants are graded in some schools according to the appropriateness of their qualifications and may or may not be encouraged by the school to undertake further qualifications. Most schools encourage teaching assistants to belong to a union; all are CRB checked before being allowed to work with students.

Associate Teachers

Associate teachers are teachers in training. Associate teachers will spend various amounts of time in schools completing their teaching practice placements in partnerships with Initial Teacher Training that is usually linked to a university. Some schools have associate teachers who are taking part on the Graduate Teacher scheme; these teachers will be in school for a much longer placement. In all cases, associate teachers are CRB checked and should be aware of current ideas in geography. Hopefully they will be enthusiastic and willing to learn about all students.

Involving older students

Some schools use sixth form or Year 11 students to support particular individuals or classes. Students volunteer to give up a study period to support others for a variety of reasons: Education Maintenance Allowance, UCAS or to supplement a CV tend to be the main reasons. Clearly staff need to use their discretion in deciding the appropriateness of pairing up students to lower school classes.

There are, of course, differences in what you can expect these people to do. Teaching assistants are working under your direction and will happily use their initiative if supported in so doing. They probably know the students better than you do, as they see them in a variety of lessons and contexts. They may be able to advise you as to what strategies work with which students. Teaching assistants are a mine of useful information; remember, they see many different teachers tackling that same student. They know too that both you and the students have good and bad days. They are not there to judge.

Associate teachers have more time to think about and develop effective geography resources, they can afford the attention to detail that we would ideally love if other pressures from other classes or other students weren't there. Associate teachers can observe and reflect as events happen when you are teaching and as such report back to you. Ideally, teacher and associate teacher learn from each other. Just remember how you impressed your main practice department with that new approach to industrial location!

Many schools have a structured system to support Year 11 and/or sixth form students in developing their sense of community. Often this involves organised and supported work with younger students. If your school currently doesn't offer such opportunities, you may like to suggest that the staff discuss the benefits and issues that such schemes present. Often the students re-learn curriculum content, as well as developing their social awareness through such experiences. Students can help you free up your time to get round to those individuals you always would wish to spend more time with. When directed appropriately to those who always finish first, those who need the security of checking each stage with someone else or those who crave praise, older students can help you keep the number of waiting hands down.

There will always be other members of staff who seem to subscribe to the school of thought that any other adult in their classroom is an intruder. However SENCOs tell us that geography staff who have read the SEN information and who have had support from teaching assistants in the past are now banging down their doors and bribing their teams with biscuits to become time-tabled support in geography.

Practicalities of managing and supporting

This chapter will now focus on managing teaching assistant (TA) support. It is often, though not universally the case, that TAs lend their expertise to teachers in terms of what works with whom, and teachers lend their expertise to TAs in terms of geography content. Working together usually means that teachers manage and support TAs and TAs manage and support teachers. Gone are the unwelcome notions of a 'them', the teaching assistant, and 'us', the teachers, but to promote positive learning they become a 'teaching team'. This is explored in more detail below, and is on the CD.

Achieving Good Practice

1. The basics

- Introduce your teaching assistant and promote them as a member of staff. The students will see you as a team and this raises standards of both good behaviour and learning.
- Foster student belief that the teaching assistant is a valued member of staff who deserves co-operation and respect. Teaching assistants need to feel confident and valued in your room to work at their best.
- Ensure you never leave your teaching assistant in an uncomfortable situation – arrive on time, never ask them to supervise classes alone – they may well do but that's your job! Although the students should see you as a team, teachers are ultimately in charge.
- Establish and negotiate with your teaching assistant working practice – is it acceptable to you for them to continue writing when you have asked the students to put their pens down? Nobody likes a dictator who barks required behaviour, so 'negotiate' is the key word here.

2. Developing a workable relationship

- Teaching assistants need to know what they are doing and where you are up to. You may not have a teaching assistant in each lesson with a particular class and it may have been two years since a particular teaching assistant did any geography. Provide your teaching assistant with copies of schemes of work to help them appreciate the aims and objectives of what you are doing.
- Decide if it is necessary to meet with them before the lesson, each week, each month to go over your plans. Is there any information that your teaching assistant needs to know before the lesson? Are there any special instructions?
- Teaching assistants find copies of answers, good pieces of work etc. useful. Everyone needs support.
- Tell teaching assistants exactly what you want them to do, with whom, for how long and when.
- Let your teaching assistant know if there is a problem, something that drives you mad. Be tactful and you won't go far wrong. Give them opportunities to respond and approach you if necessary.

It may also be helpful to make time for the teacher and teaching assistant to review curriculum planning together. On occasions you may negotiate with your teaching assistant for them to review how they thought the lesson went . . . particularly if you are using a supportive teaching strategy for the first time.

A Teaching Assistant's place in the world

There has been a lot written and broadcast recently about the current role and planned future role of teaching assistants, known also as learning support assistants or statement support assistants in other authorities. Roles and responsibilities have become increasingly specific and focused and the skill set and qualities demanded of teaching assistants at interview is ever increasing and demanding.

Job descriptions and levels aside all assistants are in schools to work with students with Special Educational Needs. Different schools organise that support in different ways and allocate that support according to different criteria. You should familiarise yourself with how your school allocates its support. This will enable you to have the best chance of receiving some support. SENCOs are directed to support students with statements first; most schools then allocate support to students at School Action Plus. Any hours left after that can be negotiated but SENCOs will timetable teaching assistants where they know they will be used and valued.

There are many ways of deploying teaching assistants. In some schools they have found it more effective when a TA follows a subject area rather than a student. This has two very positive aspects. Students do not become reliant on one person but benefit from several different approaches to learning and are not disturbed by possible absences of their own personal teaching assistant. The teaching assistant also benefits by becoming quite knowledgeable in their subject area. The teaching assistant also becomes familiar with the schemes of work and marking schemes and because of this continuity supply staff are better supported.

Relationships that are built with the teaching assistant benefit the teacher greatly as the teaching assistant can give valuable insights and tips on differentiation. Often these will not only benefit the student with the support but also the rest of the class. If the teaching assistant is not assigned specifically to a student it gives the teacher more opportunity to be creative with their deployment. In some instances it might be more productive to use the teaching assistant with the more able students to allow the teacher time to reinforce and develop major teaching points with others.

It is also important for newly qualified teachers to be introduced to the teaching assistants as they can give them a snapshot of the students they will be teaching. This is the first step in team building. It raises the profile of the teaching assistant and confidence increases in the abilities of both the teacher and teaching assistant.

Teaching assistants work to promote inclusion by:

- working under the direction of the class teacher when in the classroom, supporting one individual or a small group of individuals at either School Action, School Action Plus or Statement
- working under the direction of the SENCO when working out of the classroom with any student; on the SEN Register, following a specific programme or scheme tailored to a small group or individuals needs
- having knowledge of the involvement specialist agencies may have with a student
- contributing to the planning, delivery and review of students' IEPs by keeping detailed records
- identifying specific needs of other students within the class and alert the class teacher and SENCO

Some teaching assistants may:

- help with the preparation of resources for a particular student
- further differentiate resources to meet the needs of individuals
- organise specialist support and equipment for particular students in particular subjects
- act as a mentor to students in a pastoral sense

Teaching Assistants do not replace the teacher and so:

- should not be abandoned to take care of all the students with SEN
- are not responsible for planning and delivering and completing IEPs or any other paperwork associated with students with SEN (although as their confidence and skills develop they are an invaluable asset during this process) One important note to make at this point is that some teachers confuse teaching support with administrative support

Below is a list of the most common teaching assistant activities:

- scribing
- repeating and simplifying instructions
- prompting and encouraging
- organising a student or their work
- providing formats for students to follow
- modifying resources or tasks for students with sensory or physical disabilities

These points are not meant to be exhaustive. You should check your school's SEN policy and your department's policy on using teaching assistants for further specific information. These points are on the CD.

Each student is different and their needs differ. What a teaching assistant does for one student with a visual impairment is not the same as what they deliver to a different student with a visual impairment, as illustrated in the example below.

Example

Ryan is in Year 7. Chris his teaching assistant, ensures he carries spare board pens with him in case the teacher runs out. She checks he has his sloping board each day and goes with him once a term to see his Visual Impairment Advisory Teacher. In geography, Chris ensures Ryan has access to his CCTV enlarger, which she books out from the SEN department particularly for image or map work. Katie is in Year 11 and also has a visual impairment. She needs all work enlarging from A4 to A3. Chris accompanies Katie on all fieldwork, both inside the school grounds and on field trips away from school for health and safety reasons.

How might support work in a geography classroom?

INTRODUCTORY SUPPORT	
The teacher could support the assistant and student by	**The assistant could support the teacher and student by**
Providing the lesson plan/scheme of work and any necessary answers highlighting TA role	Reading the relevant chapter or page, glance over good examples of finished work
Communicating any instructions, allowances or expectations to be made for particular students	Looking at the resources the teacher intends to use to check for suitability
Preparing or differentiating further any resources to free up TA time in class	Enlarging work for visually impaired students, photocopying resources onto coloured paper for dyslexic students

For example, during a simple introductory activity for a group of Year 7 students studying volcanoes the TA proof read what the teacher had used the year before and tweaked it for the small group of students with numeracy difficulties in the group. The students had to decide which 10 things they would take from their house to the emergency shelter once the eruption warning siren was heard. Each of the 25 options had a weight and their rucksack could carry 20 kgs. The TA simplified the weights to multiples of 2, 5 and 10 and checked that the students would have the vocabulary knowledge to choose from all 25 options.

At the start of a geography lesson

INITIAL LEARNING	
The teacher could support the assistant and student by	**The assistant could support the teacher and student by**
Indicating the enquiry process to be followed on lesson plans	Asking leading questions, draw in reticent students
Telling the TA who needs what in which order through the lesson	Helping teacher manage tasks by issuing resources
Telling a particular student that Mr X or Miss Y will be checking on them	Re-focusing a student's attention
Setting agreed time checks within the lessons	Giving feedback to teacher about who has/hasn't finished and common stumbling blocks
Providing class list or list of students for particular assessment	Observing and recording contributions on class list for teacher

For example, when analysing photographs of Nairobi, the teaching assistant produced a laminated spider diagram with a spinner on it. The spinner was spun and it stopped on one of 10 key geographical enquiry questions to help the students analyse the image.

During the main teaching period

DURING THE MAIN LEARNING	
The teacher could support the assistant and student by	**The assistant could support the teacher and student by**
Providing the TA beforehand with the opposing arguments or main points in discussion work	Enacting the other viewpoint in a role play, play devil's advocate in a debate
Ensuring the TA knows where pens, OHPs, plugs, spare paper etc. is kept to ease whole class disruption	Acting as a scribe in a student's book, or for the teacher on an OHP or white-board when students volunteer answers
Handing out answers to the TA beforehand, highlighting any difficult or tricky questions or exceptions to the rule	Pre-empting and trouble shooting common mistakes or errors
Ensuring the TA knows aims and objectives for the lesson, provide them with their own copies of handouts etc.	Re-directing reading, encouraging and supporting individuals
Telling the TA what you want them to check: geographical content? Spellings? Use of terminology and key words? IEP target?	Helping students check their work
Ensuring the TAs know what has to be achieved for the majority and any allowances to make for individuals	Praising students, handing out extension work. Questioning students about what they have learnt for the plenary

For example during a lesson on changing employment structure the TA recorded the jobs that the student's parents did on the OHP. She recorded them in three different colours for primary, secondary and tertiary to help the class work out the differences. Later during the write up she used the teacher's simplified flow diagram to help the students classify a list of other jobs and then volunteered her own family example to show how employment opportunities have changed over time in Stoke on Trent.

Towards the end and after the lesson

CLOSING THOUGHTS	
The teacher could support the assistant and student by	**The assistant could support the teacher and student by**
Setting up expectations that all students will contribute at some point if able	Prompting the students to supply answers, praising and encouraging
Ensuring the student believes good work will be recognised with merits or credits or positive comments	Mentioning the good work a particular student has done
Ensuring homework is not set quickly in the last minute of the lesson	Noting down the homework for the student, checking they know what to do
Asking for the TA's views on progress	Informing the teacher of progress
Ensuring the TA feels confident and valued to comment	Being involved in evaluating this lesson and planning and preparing resources for the next

For example the teaching assistant suggested that a writing frame that gave more support to David might be needed to help him extend his writing. The class had been studying migration in Italy. At the moment he could only think of the effects on those already living in Turin when discussing Guiseppe's move from Aliano. He had not grasped the consequences for Guiseppe's own town. David had used the frame well to write about two effects on the people and two on the environment but he had not made any other links.

The summary guidelines on the sharing of responsibilities of the teaching team given on the CD can usefully be used to structure a departmental session on the effective and purposeful use of support staff. Both teachers and teacher assistants could annotate this list with geography-specific examples as to how they can modify their work to reduce the barriers to learning in geography.

Clearly a teaching assistant who takes part in many lessons per week will need advance warning and time to prepare for some of the points made in the guidelines on the CD. In addition the teacher will need to make extra preparations for the teaching assistant to support them effectively.

Teaching assistants are not there to do general photocopying or other types of administration.

How can support help break down barriers to learning?

Accommodating a range of learning needs in a classroom situation and maintaining contact with the curriculum is a problem ever present in every subject area in secondary education.

This Year 8 group at Moorside High School, Werrington, is typical. It is a smaller set than average, 18 students to be exact, all with different barriers to learning. Within this class there are five students with statements, one looked after child, one school refuser and one with temporary visual impairment.

The challenge of the class teacher is to engage all the students and the more able and enable them to achieve and aspire.

- Adam is a severe Dyspraxic (DCD – Developmental Co-ordination Disorder) with the problems this 'label' encompasses (see pages 32–34 in Chapter 3).
- Ben, Carl and Donna all have specific learning difficulties (see pages 27–34 in Chapter 3).

Ben and Carl demonstrate a higher level of performance orally and participate in all lessons. Donna, on the other hand, has poor self-esteem and questions have to be directed as she rarely participates. They all have problems in gaining literacy and numeracy skills having:

- difficulty with sequencing
- difficulty with visual perception
- difficulty in working memory
- delays in language function.

The impact of this is that they rarely demonstrate their true capabilities in written work.

Gareth is another member of the class. His problems are extremely complex. He has been diagnosed as having Pathological Demand Avoidance Syndrome and Pragmatic Language Disorder (see pages 52 to 55 in Chapter 3 on Communication and Interaction). He spent his formative years in a special school due to his inability to communicate verbally. To his and his parents' credit he was able to transfer to mainstream primary school when he was in Year 5. Gareth is an able student, whose difficulties lie in his obsessive behaviour and his inability to understand language and its inference. Everything is literal to Gareth.

If you ask Gareth if he has a pen when meaning you wish him to use it to begin his work, he will quite readily respond with 'Yes, thank you!' but still fail to commence writing. He is extremely adept at diverting your attention and engaging you in conversation on a topic of his choosing (not the lesson objective), thus avoiding the task he does not wish to do.

Due to the range and diversity of the needs this Year 8 class presented, extra time was set aside for planning to try and pre-empt any problems that might affect the learning of the class. The teacher and the teaching assistant worked together on the indicators of the students' problems and strategies to address them. The following lesson was planned in advance and ultimately delivered by teacher and teacher assistant.

Classroom tips for a diverse range of abilities

- Coloured paper
- Handouts (keep writing on the board minimal)
- Stage instructions and keep them simple (too much information can confuse)
- Direct the TA to the area where instruction needs reinforcing and repeating

The lesson focus was to learn about and develop an understanding of the Water Cycle. Appendix 7.1 was used. This also appears on the CD and it includes sentences to be cut up into cards to sequence. The key words for this sequence of learning were identified.

Evaporation
Condensation
Cycle
Extension – Weathering
Erosion
Skills – Identify a logical sequence.

The teacher communicated to the teaching assistant that as a minimum he expected all students to know and understand the key words. He expected the majority to be able to sequence the events of the water cycle correctly and that some students may access some extension activities relating to erosion and weathering. A few students may start to make connections between weathering, erosion and the water cycle. It was agreed that the TA would work in the following ways to break down the barriers to learning. For the students exhibiting characteristics of dyspraxia, she would facilitate the following support:

- put the cards in an envelope
- look in dictionary for definition of key words
- support the student to use laptop to write down definition
- support the student to put cards from envelope in logical sequence
- support student to copy on to laptop
- remind the student to carry laptop

For the students with specific learning difficulties, the teaching assistant would work with them to:

- look up key words in the dictionary
- use paired reading to access information on the cards in the envelope
- use information from the cards and the picture as a base to illustrate the water cycle. (Experience and prior knowledge of the students have lead the teaching assistant and teacher to use this strategy)
- become familiar with the key words

For the students with communication and interaction needs the teaching assistant:

- reinforced the tasks to be completed
- used the envelope and numbered each card to reinforce sequence
- encouraged a specific number of sentences to be written. (Gareth will write reams of text in any order including information he needs)
- focused attention on key points and set a limit of number of sentences, i.e. five sentences on the water cycle

Other general strategies that were used with the class included a partially completed flowchart, and the use of mind mapping (Mind Manager Smart I.T.).

As a general rule not all classes have the diversity of problems as this Year 8 class. In most cases at Moorside High School a daily support observation record is filled in by the teaching assistant for each class. (Appendix 7.2 and on CD.) This is a simple tick chart with the names of students given support. This helps identify two areas. Firstly, areas where the amount of extra help is needed in specific topics. This is an extremely good indication as to whether the lesson has been pitched at the correct level. Too much help – perhaps the topic needs to be approached from a different angle. Too little help – perhaps the topic is not challenging enough. Secondly, the chart identifies both areas of extra need with the students with statements and those students who perhaps are struggling unnoticed. Gathered over several lessons, these charts can be evaluated and subject specific targets can be set.

A second report chart (Appendix 7.3 and on CD) is also a valuable asset when attempting to address persistent disruptive behaviour. These sheets are filled in by the teacher after consultation with the teaching assistant as to particular concerns. The form does not only target these concerns but strategies that work and that can be shared by various members of staff who teach the same student. Once again the information compiled can be reused to write specific targets for Individual Behaviour Plans.

Meeting Special Needs in Geography

Summary

Working effectively with colleagues to plan and implement an appropriate geography curriculum can have very positive benefits to learning for all, adults and students alike. Relationships are key. Respect and understanding are often strongly evident in effective geography classrooms. Purposeful planning and liaison are also features of successful learning environments. Teachers are encouraged to think with clarity about the key learning objectives of the lesson. Such quality curriculum thinking is of benefit to all students, not just those with specific identified needs. The teaching assistant is able to support students in developing their understanding and sense of the world from an individual perspective.

Appendices

Appendix 1.1
Map Symbols for an Accessibility Map

Appendix 2.1
Developing a Departmental Policy – INSET Activity

Appendix 2.2
Referral form, St. John Fisher Special Needs Department

Appendix 2.3
Praise form, St. John Fisher Special Needs Department

Appendix 2.4
SEN and Disability Act 2001 (SENDA)
INSET Activity

Appendix 3
The Opportunities and Challenges of a Geographical Education for Students with Special Needs

Appendix 4.1
How Geography Contributes to our Mission Statement
and School Aims at Stretton Brook School

Appendix 4.2
Using a Sensory Room to Develop a Sense of Place

Appendix 5.1
Local Leisure – Wheelchair Enquiry

Appendix 5.2
Strategies to Reduce Barriers to Outdoor Learning

Appendix 5.3
Tourism Word Mat

Appendix 6.1
A Glossary of the Principle Features of Educational Assessment

Appendix 6.2
Learning Activities for Assessment

Appendix 6.3
Student Friendly P Levels for Geography

Appendix 7.1
Water Cycle Cards

Appendix 7.2
Daily Support Observation Record

Appendix 7.3
Report Chart

Map Symbols for an Accessibility Map

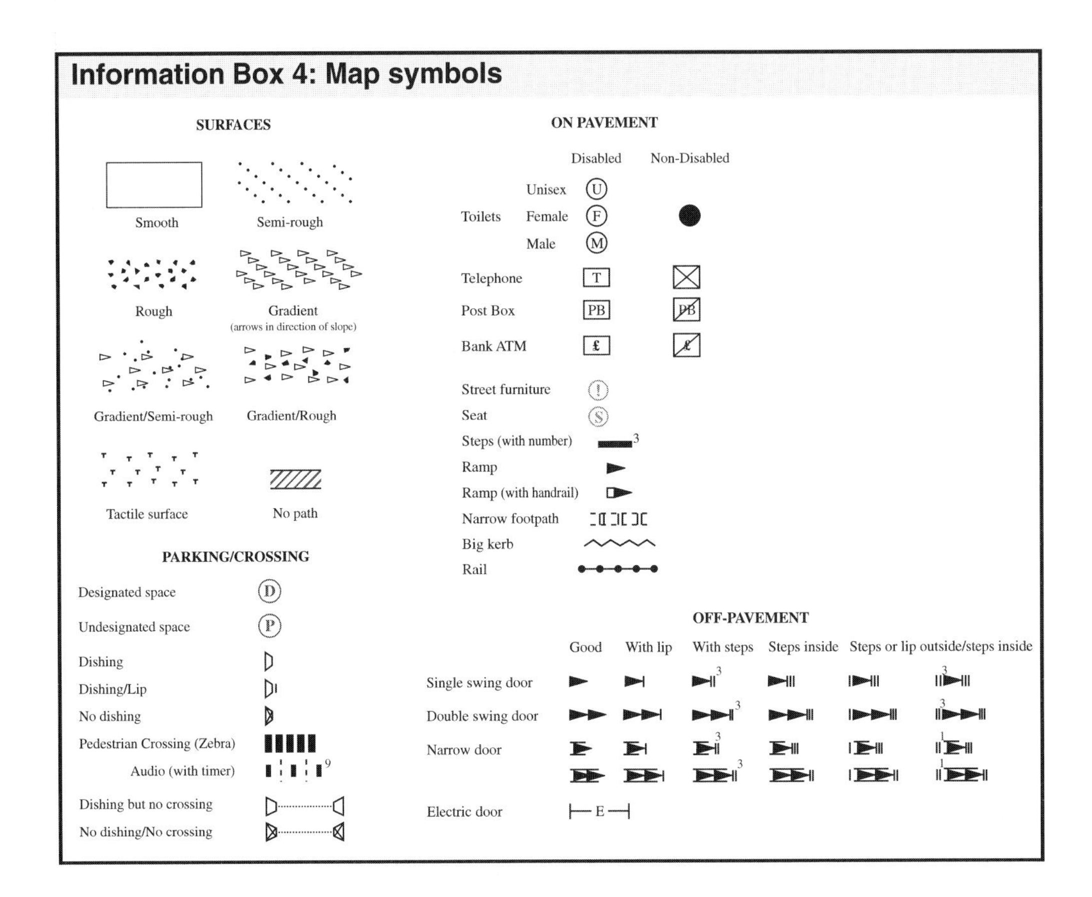

Developing a Departmental Policy INSET Activity

What do we really think?

Each member of the department should choose two of these statements and pin them on to the noticeboard for an overview of staff opinion. The person leading the session (Head of Department, SENCO, senior manager) should be ready to address any negative feedback and take forward the department in a positive approach.

If my own child had special needs, I would want her/him to be in a mainstream school mixing with all sorts of kids.

I want to be able to cater for students with SEN but feel that I don't have the expertise required.

Special needs kids in mainstream schools are all right up to a point, but I didn't sign up for dealing with the more severe problems – they should be in special schools.

It is the SENCO's responsibility to look out for these students with SEN – with help from support teachers.

Students with special needs should be catered for the same as any others. Teachers can't pick and choose the students they want to teach.

I need much more time to plan if students with SEN are going to be coming to my lessons.

Big schools are just not the right places for blind or deaf kids, or those in wheelchairs.

I would welcome more training on how to provide for students with SEN in geography.

I have enough to do without worrying about kids who can't read or write.

If their behaviour distracts other students in any way, youngsters with SEN should be withdrawn from the class.

ST. JOHN FISHER SPECIAL NEEDS DEPARTMENT STUDENT REFERRAL FORM*

Teacher: .... Date ....

Department: ....

Name of Student:.. Form.......... Class.............

NATURE OF CONCERN COGNITION AND LEARNING	BEHAVIOURAL/EMOTIONAL
☐ Reading fluency/speed	☐ Poor concentration
☐ Reading accuracy	☐ Hyperactivity
☐ Spelling /Grammar	☐ Emotional immaturity
☐ Handwriting speed	☐ Maladjustment to expectations
☐ Handwriting legibility	☐ Poor interaction with peers/adults
☐ Other	☐ Other

COMMUNICATION/INTERACTION	SENSORY/PHYSICAL
☐ Expression	☐ Hearing
☐ Articulation	☐ Eyesight/Vision
☐ Attention span	☐ Mobility
☐ Frustration	☐ Other
☐ Other	

Evidence for Referral

Severity of concern --

Disadvantages/Problems concern is creating for student/class/group

----

Details of interventions already tried in your classroom to deal with concern

----

Any other comments

Action undertaken in response to referral

----

* Please see CD for enlarged copy

ST. JOHN FISHER SPECIAL NEEDS DEPARTMENT STUDENT PRAISE FORM

Student Name

Teacher

Class

Subject

Form

Which IEP target is the praise concerning?

Why does the student deserve praise?

Any other comments?

SEN and Disability Act 2001 (SENDA) INSET Activity

1 The SEN and Disability Act 2001 amends the Disability Discrimination Act 1995 to include schools' and LEAs' responsibility to provide for pupils and students with disabilities.

2 The definition of a disability in this Act is,

'someone who has a physical or mental impairment that has an effect on his or her ability to carry out normal day to day activities. The effect must be:

- substantial (that is more than minor or trivial); and
- long term (that is, has lasted or is likely to last for at least a year or for the rest of the life of the person affected); and
- adverse.'

Activity: List any pupils that you come across that would fall into this category.

3 The act states that the responsible body for a school must take such steps as it is reasonable to take to ensure that disabled pupils and disabled prospective pupils are not placed at substantial disadvantage in comparison with those who are not disabled.

Activity: Give an example of something which might be considered 'a substantial disadvantage'.

4 The duty on the school to make reasonable adjustments is anticipatory. This means that a school should not wait until a disabled pupil seeks admission to consider what adjustments it might make generally to meet the needs of disabled pupils.

Activity: Think of two reasonable adjustments that could be made in your school/ department.

5 The school has a duty to plan strategically for increasing access to the school education, this includes provision of information for pupils and parents (e.g. Braille or taped versions of brochures) improving the physical environment for disabled pupils and increasing access to the curriculum by further differentiation.

Activity: Consider ways of increasing access to the school for a pupil requesting admission who has Down's Syndrome with low levels of literacy and a heart condition that affects strenuous physical activity.

6 Schools need to be proactive in seeking out information about a pupil's disability (by establishing good relationships with parents and carers, asking about disabilities during admission interviews etc.) and ensuring that all staff who might come across the pupil are aware of the pupil's disability. This will include meaningful liaison between schools.

Activity: List the opportunities that occur in your school for staff to gain information about disabled students. How can these be improved upon?

The Opportunities and Challenges of a Geographical Education for Pupils with Special Needs*

Opportunities to be celebrated and built upon include	**Challenges to be faced and reduced to support learning**
The opportunity to use real world examples both locally and at a distance	The challenge of using appropriate support strategies to access maps
The opportunity to include and build upon the pupils' own experiences	The challenge of accessing information in an atlas
The opportunity to encourage freedom of movement by developing a confidence about places and spaces	The challenge of what not to do and what to emphasise
The opportunity to develop a sense of the pupils' own space and place	The challenge of accessing appropriately outdoor learning
The opportunity to develop an awareness of different environments and how we use and mis-use them	The challenge of making a variety of visual resources about a place accessible
The opportunity to be curious about places and spaces	The challenge of supporting the issue of specialist vocabulary
The opportunity to develop an awareness of others, other people, other places	The challenge of making the pupil comfortable so that they can learn effectively, both indoors and outside
The opportunity to consider real people and how places both enable and disable them	The challenge of creating multi-sensory activities, including sensory trails of regularly visited places
The opportunity to use a variety of place, theme or concept based resources, e.g. artefacts, objects, pictures, photos	The challenge of using real world rather than abstract examples of places and processes
The opportunity to learn in the real world through outdoor experiences	The challenge of creating a resource bank, e.g. taped descriptions of regularly used maps and photos
The opportunity to focus on the visual through the use of images	The challenge of making certain that pupils know how tasks clearly relate to the learning objectives
The opportunity to create many sensory experiences related to place	The challenge of breaking learning into small steps and making the sequence clear to the pupil
The opportunity to frame geographical questions that are initiated by the pupils	The challenge of supplementing text with visual and auditory resources and vice versa
The opportunity to understand geographical processes, e.g. water cycle	The challenge of organising social groups appropriately

* Please see CD for enlarged copy

How Geography Contributes to Our Mission Statement and School Aims at Stretton Brook School

We believe that geography is an important subject since it contributes to all aspects of the school mission statement by:

- the variation of topics studied and differing teaching methods which allow students to progress at their own rate and reach their full potential
- through learning about their environment students develop a sense of identity
- teaching activities include creative work, discussions, group work and opportunities for collaborating with others
- by learning about their own locality students become aware of the community in which they live
- geography is about people, as well as place, and through learning about different races and cultures, about world problems and environmental issues, students develop a sense of awareness and a more tolerant understanding of students outside their own experience
- stressing a positive framework which emphasises individual achievements and successes
- providing a curriculum which is relevant to our students with which they can relate, and by giving opportunities for fieldwork and visits beyond their immediate environment as well as allowing for differentiation
- enhancing self-esteem and respect which in turn encourages a desire to learn and thus promoting more positive values
- encouraging an awareness of other people from all aspects of society and from differing races and cultures and so engendering more positive attitudes to equal opportunities

Using a Sensory Room to Develop a Sense of Place

The sensory room – a resource for geography and literacy

Adapted from ideas originally placed by Chris Durbin on the Staffordshire Learning Net, www.sln.org.uk/geography in 2002.

If geography is about increasing awareness of the wider world in all its diversity and beauty, then students with severe learning difficulties should have an opportunity to experience a range of different places at first hand. This sensory approach and the sensory room in particular is an opportunity to provide these opportunities second hand.

The sensory room can be set up to represent a place

For example, it could be:

- A tropical rainforest
- The seaside on a stormy day
- A Caribbean banana plantation
- An African market
- The middle of a big city

What you will need

- Either a set of slides and a slide projector, or a computer and data projector with images from the web projected on the wall.
- The room heated to the approximate temperature (and humidity) of the place (e.g. kettles boiled in the room to create tropical rainforest heat and humidity).
- The floor dressed with appropriate material (e.g. sand and pebbles for a beach).
- Some sounds of the place – recorded on a portable tape recorder – music can be appropriate too but you can download sound effects from the web or buy them on CD or tape (e.g. the sounds of a street in a big city).
- Smells recreated by cooking food or spraying aromas (e.g. the smell of fruit and vegetables of an African market). These can be tasted if appropriate too.
- Artefacts that can be handled, if appropriate (e.g. rocks, seaweed, driftwood from the beach).

The key is to think how the environment you are studying can be recreated.

Local Leisure – Wheelchair Enquiry

Lesson Plan, Bev Rowley, Stretton Brook School

In which ways are the leisure facilities of Burton upon Trent more accessible to some groups rather than others?

Creating a Need to Know

Students will be divided into groups. Each group will be given a wheelchair. They will be asked to move around school and find areas that are inaccessible to wheelchair users. Following this activity in the next lesson students will work in groups. One member will be disabled for the afternoon. Using a wheelchair they will visit the leisure facilities of Burton on Trent to investigate real and perceived barriers to their freedom of movement.

Initial Response

Students will discuss:

- What is leisure?
- Does leisure and its use vary to meet individual needs?
- Do leisure facilities vary in the way that they meet the needs of people with lesser degrees of mobility?
- Are all leisure facilities accessible to all?

Research using data

Research related questions include:

- How can students find out more about this?
- Who could they present their findings to?
- How will they present their findings?

Students will have already conducted an investigation of leisure facilities in Burton upon Trent and how leisure and its use varies according to:

- Individual needs
- Individual age
- Degree of mobility

- Accessibility of amenities
- Individual's leisure time
- Economic and social conditions

Students will brainstorm their existing knowledge of the towns leisure facilities and search for other information using:

- Leaflets
- Tourist information
- Internet

Students will mark various places on a large street map of Burton on Trent.

Making Sense

Students will visit various leisure facilities and will gather information about it through:

- Leaflets
- Personal experience of being disabled
- Telephoning places/interviewing staff

They will produce their own leaflet for each facility describing which groups this activity will appeal to, its costs, whether it is on a bus route etc.

Next to this they will rate each facility according to its accessibility to wheelchair users. Photographs that have been annotated will be used.

Written or tape-transcribed accounts of how each of the wheelchair users felt throughout this activity will also be on display.

The wall display will show ratings of how accessible each place is for wheelchair users. The purpose of this is to increase awareness of barriers and how these may restrict freedom of movement and choice for individuals.

Reflecting on Learning

Students' local locational knowledge and mapping skills will be assessed through their ability to find places on a street map.

Students' leaflets of leisure facilities and their accessibility ratings will be assessed.

Students will be given verbal feedback about the strengths and weaknesses of their investigation.

Strategies to Reduce Barriers to Outdoor Learning

Barriers to Learning and Strategies to consider using	Dyslexia See pages 27–9	Dyscalculia See pages 30–2	Dyspraxia See pages 32–4	MLD See pages 34–6	SLD See pages 37–9	PMLD See pages 39–42	Fragile X See pages 42–4	Down's Syndrome See pages 44–6	BESD See pages 47–9	ADD/ADHD See pages 49–51	Comm. Interaction See pages 52–5	ASD See pages 55–7	Asperger's See pages 57–60	VI See pages 61–4	HI See pages 64–6	Multi-Sensory See pages 66–8	PD incl. Cerebral Palsy See pages 68–71	Tourette's See pages 71–3
Staff can pre-read handouts to students	x													x	x			
Tasks can be given out before the visit					x									x	x			
Use digital images/video to create a visual narrative about the place that can be used both pre- and post-visit					x	x					x	x	x	x	x	x		
A sensory memory box could be created						x								x		x		
Use clear line sketches to emphasise information provided on maps and photos				x	x									x				
Make tape recordings of descriptions on maps/photographs											x			x		x		
A tactile/sensory trail can be developed, linked to the planned activities					x	x								x		x		
Recording sheets can be photocopied onto coloured paper	x																	
Recording sheets can be adapted		x																
Recording sheets can be structured using writing frames and cloze procedures			x	x				x										
Recording sheets – clearly link the activity to its purpose							x		x	x	x	x	x					

Barriers to learning and Strategies to consider using	Dyslexia See pages 27–9	Dyscalculia See pages 30–2	Dyspraxia See pages 32–4	MLD See pages 34–6	SLD See pages 37–9	PMLD See pages 39–42	Fragile X See pages 42–4	Down's Syndrome See pages 44–6	BESD See pages 47–9	ADD/ADHD See pages 49–51	Comm. Interaction See pages 52–5	ASD See pages 55–7	Asperger's See pages 57–60	VI See pages 61–4	HI See pages 64–6	Multi-Sensory See pages 66–8	PD incl. Cerebral Palsy See pages 68–71	Tourette's See pages 71–3
Recording sheets – think about complexity of diagrams, which information is essential?				x	x									x				
Consider preferred communicated system for outdoor learning						x									x		x	
Appropriate pairs/groups of students can be created to support each other	x	x		x			x	x		x	x	x	x	x	x			x
Additional visual information can be used, e.g. photos, line drawings, symbols		x	x	x										x				
A clear description of the sequence of events can be given			x				x	x	x		x	x	x	x				
A clear description of how the day is organised				x			x	x	x		x	x	x		x			
Communicate throughout the visit what is happening and what will happen next						x					x	x	x					
A pre-visit could be arranged				x	x													
The terrain should be reviewed to evaluate accessibility			x			x		x						x				
There are clear guidelines given in an appropriate format about expected behaviour				x		x	x	x	x	x	x	x	x					x
Plan for appropriate adult support						x			x	x								x
Provide portable schedule and work system						x												
Identify and communicate appropriately potential dangers and strategies that can be drawn upon to deal with them									x	x				x	x			

Tourism Word Mat

PLACES

Peak District
Disneyland Paris
France
Greece
Kenya
Mediterranean
Majorca
Menorca
Spain
Turkey
USA

KEYWORDS

activity holiday
advertising
attractions
airport
beach
biodiversity
brochure
climate
conflict
conservation
destination
ecotourism
honeypot
hotel
images
national park
package holiday
sightseeing
skiing
sustainability
tourism

ADJECTIVES

attractive
brash
coastal
cultural
historic
lively
mountainous
natural
quaint
quiet
scenic
spectacular
sunny
tropical
vibrant
wild

PROBLEMS

seasonal jobs
hotels owned by foreign owners
traffic congestion
increase in crime
woodlands cleared for development
noise
water shortages
increase in prices
visual pollution

BENEFITS

new jobs
investment
improved roads
foreign income
sustainable tourism
better services
new roads and airports

Thanks to T Purcell from The Grange School, Stourbridge and www.sln.org.uk/geography

A Glossary of the Principal Features of Educational Assessment

Source: Lambert, 1996 p. 261 'Assessing Pupil Attainment' in Kent, A, Lambert, D, Naish M, and Slater, F (eds) *Geography in Education, Viewpoints on Teaching and Learning*. Cambridge: Cambridge University Press.

Feature	Meaning	Feature	Meaning
Formative assessment	Assessment to support future learning during a course of study	Summative assessment	Assessment undertaken at the end of a course of study
Formal assessment	Includes a degree of standardised procedures, as in tests	Informal assessment	Based on observation and conversation with pupils
Formal Records	Often numerical, consisting of marks, grades, etc.	Informal records	Qualitative information, often carried in teachers' heads
Marking pupils' work	One part of the overall knowledge building process; in practice can be little more than monitoring work done	Criteria referencing	Pupils' work is judged in relation to explicit criteria which identify progress by describing levels of attainment
Norm referencing	Pupils' work is judged in comparison with the performance of other pupils	Ipsative assessment	Pupils' work is judged solely in the context of the individual pupil's previous performance and circumstances
Validity in educational assessment	Usually refers to the content or strategy adopted: is this assessing what I think it is assessing?	Reliability of assessment	Usually a reference to the influence of external factors on outcomes: how well standardised are the questions, procedures and marking?
Fitness for purpose	Assessment information has several purposes to which it may be put; does the adopted assessment method provide data in the right form?	Achievement	A broader concept than attainment; includes non-academic goals such as motivation, social and personal skills
Attainment	Usually described as a level in relation to specified attainment targets	Teacher assessment National Curriculum	A summative judgement made near the end of the Key Stage based upon the pupil's overall performance and progress
Performance	A range of tasks, exercises, etc. provide the evidence on which judgements relating to attainment are based	Ability	A complex idea; over-hasty extrapolations about general 'ability' on the basis of limited evidence drawn from pupils' performances are best avoided

Learning Activities for Assessment

Adapted from p. 275 of Capel, S, Leask, M, and Turner, T (1995) *Learning to Teach in the Secondary School: A Companion to School Experience.* London: Routledge

Oral Evidence	Written Evidence	Graphic Evidence	Products
Questioning Listening Discussing Presentations Interviews Debates Audio recording Video recording Role play Simulation	Questionnaires Diaries Reports Essays Notes Stories Newspaper articles Scripts Short answers to questions Lists Poems Descriptions PowerPoint presentations	Diagrams Sketches Drawings Graphs Printouts Overlays Maps Annotated photographs Storyboards Animations Geographical information systems	Models Artefacts Games Videos Photographs Recordings

Consider

- Which of these are produced frequently in your classroom?
- Which are produced infrequently?
- Why?
- Can you add to this list?

Student Friendly P Levels for Geography

Level	Description	Achievement
P1	I can experience geographical activities I can simply respond by noise/actions to sounds/environmental changes, e.g. trees in wind	
P2	I can recognise my family/familiar objects I can enjoy touching/experiencing different textures	
P3	I can communicate that I am enjoying a geographical activity/experience I can explore different materials, e.g. sand, water, pebbles I can remember simple learned responses, e.g. plant a plant	
P4	I can know familiar places locally, e.g. park, school I can use simple words about the world	
P5	I can follow set rules about a familiar place I can answer orally a question about places and people	
P6	I can understand differences between natural and human made features I can begin to sort objects in terms of simple features, e.g. colour, size, where found etc. I can use symbols to make simple maps	
P7	I can use phrases about preferences about natural/human made places I can show directions by using symbols and models I can show simple care for my own environment	
P8	I can begin to use common words for locations I can use symbols on maps and plans I can show some understanding about environmental issues I can answer simple questions about buildings and their uses	

Water Cycle Cards

The hot air containing this water vapour rises and cools	These clouds are moved towards the land by winds
Underground water slowly moves towards the oceans or appears on the surface again as a spring	The clouds rise over the land and this causes rain
The rivers run downslope to the oceans	This water later evaporates back into the atmosphere by transpiration
The sun's energy heats any water surface, e.g. oceans, and causes the water to evaporate	The water in the oceans is again evaporated so the cycle begins once more
The rain that falls onto the ground either soaks into the soil and rocks or flows over the surface	As the rain falls towards the earth's surface, some of it is intercepted (caught) by vegetation
Water on the surface runs downhill as surface run-off and eventually flows as rivers	As it cools the water vapour condenses to form clouds

DAILY SUPPORT OBSERVATION RECORD

Moorside High School

SUPPORT TEACHER:						CLASS TEACHER:				
SUBJECT:	TUTOR GROUP:						DATE:			
	STUDENT HELP GIVEN									
NAMES OF STUDENTS HELPED	Discussion/Explanation	Organisation of class equipment	Personal organisation	Reading	Writing	Simplification of task set	Content reinforcement	Encouraged/Reassured	Concentration	Homework

OTHER COMMENTS:

* Copy also included on CD

REPORT CHART

Moorside High School

SUBJECT TEACHER'S REPORT			
STUDENT'S NAME:		FORM/SET:	
A CONCERNS			
1 Do you have any concerns about the above student? Yes No (if no complete part B only)			
2 List your two main concerns about this student – state what you can see or hear the student doing which causes you concern (e.g. 'he shouts out' rather than 'he is disruptive' or 'he hits other students' rather than 'he is aggressive'). i. ii.			
3 Frequency When was the last time each of the above occurred (tick appropriate column)?			
	In the last week	In the last month	In the last term
Concern (i) occurred			
Concern (ii) occurred			
B POINTS FOR PLANNING			
1 What would you like the student to be doing more of?			
2 List any strategies which you have found to work with this student			
3 List anything you know which the student particularly enjoys/finds rewarding			
SIGNED:		**SUBJECT:**	
Number of periods you teach this student each week:			**DATE:**
Please return this to:			**Thank you for your help**

* Copy also included on CD